Large Deviations for Discrete-Time Processes with Averaging

Large Deviations for Discrete-Time Processes with Averaging

O.V. Gulinsky and A.Yu. Veretennikov

///VSP///
Utrecht, The Netherlands, 1993

VSP BV
P.O. Box 346
3700 AH Zeist
The Netherlands

First published in 1993

ISBN 90-6764-148-0

CIP-DATA KONINKLIJKE BIBLIOTHEEK, DEN HAAG

Gulinsky, O.V.

Large deviations for discrete-time processes with
averaging / O.V. Gulinsky and A.Yu. Veretennikov. -
Utrecht: VSP
with ref.
ISBN 90-6764-148-0 bound
NUGI 815
Subject headings: Cramér's theorem / Markov processes.

Printed in The Netherlands by Koninklijke Wöhrmann, Zutphen.

Contents

Preface

Let $\xi = (\xi_k)_{k\geq 1}$ be a sequence of independent and identically distributed (i.i.d.) random variables with a common distribution μ that are defined on a probability space (Ω, F, P). Assuming that $\int_R |x|\mu(dx) < \infty$, *the weak law of large numbers* says that

$$n^{-1}\sum_{i=1}^{n}\xi_i$$

converges as $n \to \infty$ to $m = \int_R x\mu(dx)$.

Suppose that μ has mean 0 and variance 1. Then by the classical *central limit theorem* for each $x \in R$

$$F^n(x) = \mathsf{P}\left(\frac{1}{\sqrt{n}}\sum_{i=1}^{n}\xi_i < x\right)$$

converges as $n \to \infty$ to

$$\Phi(x) = \frac{1}{\sqrt{2\pi}}\int_{-\infty}^{x} e^{-u^2/2}\, du.$$

The theory of large deviations emerged as an attempt to give an answer to the question how fast the tail of distribution F^n is approaching to the Gaussian's tails. The first general answer given by Cramér (1938) has the following form: under Cramér's condition $\mathsf{E}\exp(\lambda\xi_1) < \infty$, $\lambda > 0$, for $x \geq 0$, $x = o(\sqrt{n})$ $(n \to \infty)$

$$\frac{\mathsf{P}\left(\frac{1}{\sqrt{n}}\sum_{i=1}^{n}\xi_i > x\right)}{1-\Phi(x)} = \exp\left[\frac{x^3}{\sqrt{n}}R\left(\frac{x}{\sqrt{n}}\right)\right]\left(1 + O\left(\frac{x+1}{\sqrt{n}}\right)\right),$$

where $R(z)$ is Cramér's series constructed with the help of the semi-invariants of ξ_i.

This type of result is named 'moderate deviations' and will not be considered in our book. For related work and other references, see Ibragimov and Linnik (1971, in Russian 1965) and Petrov (1975). More recent results for sums of random not necessary independent variables are presented in Saulis and Statulevicius (1989).

The prototype for large deviations results considered in this book is Cramér's renowned theorem which deals with the 'rough' logarithmic asymptotics of the probability

$$\mathsf{P}\left(\sum_{i=1}^{n}\xi_i > na\right),$$

where $a > 0$ does not depend on n. It turns out (Cramér (1938), see also Chernoff (1952)) that under Cramér's condition the following limit takes place:

$$\lim_{n\to\infty} n^{-1}\ln \mathsf{P}\left(\sum_{i=1}^{n}\xi_i > na\right) = -I(a),$$

where the *rate function* $I(a)$ is the Legendre transform of the logarithmic moment generating function $G(\lambda) = \ln \mathsf{E} \exp(\lambda\xi_1)$:

$$I(a) = \sup_{\lambda \in R} [\lambda a - G(\lambda)].$$

In a general case the sequence $(F_n)_{n \geq 1}$ of probability measures on Borel σ-field of metric space X is said to have the large deviation property if there exists a function $I(\cdot)$ such that for each open set G in X

$$\liminf_{n \to \infty} n^{-1} \ln F_n(G) \geq - \inf_{x \in G} I(x),$$

for each closed set K in X

$$\limsup_{n \to \infty} n^{-1} \ln F_n(K) \leq - \inf_{x \in K} I(x),$$

and the level sets of the rate function $I(x)$ are compact in X.

Cramér's result was next extended by many authors. Sanov (1957) was the first to formulate and prove the large deviation property on a space of probability measures. He considered large deviations for empirical distribution

$$\mu_n(A) = n^{-1} \sum_{i=1}^{n} \mathrm{I}\ (\xi_i \in A),$$

with $A \in \mathcal{B}(R)$ (here I is indicator). This formula determines the measure-valued process with value in the space $\mathsf{M}_1(R)$ of probability measures on R. If M_1 is supplied by an appropriate metric $\rho(\alpha, \nu)$, $\alpha, \nu \in \mathsf{M}_1$, it is reasonable to inquire about logarithmic asymptotics of the probability $\mathsf{P}(\rho(\mu_n, \nu) \leq \delta)$, $\nu \in \mathsf{M}_1$, $\delta > 0$. The rate function for this problem was found to be the relative entropy function

$$I(\nu) = \int_R \ln \frac{d\nu}{d\mu}(x)\, \nu\, (dx), \qquad \nu \in \mathsf{M}_1, \qquad \nu \ll \mu,$$

introduced into statistics by Kullback and Leibler (1951). At the same time this rate function can be represented as the Legendre transform of the corresponding logarithmic moment generating function. The large deviation theory for the occupation time functional was next developed by Donsker and Varadhan (1975a, 1975b, 1976) for a Markov chains and processes. Similar results were independently obtained by Gärtner (1977).

Bahadur and Zabell (1979) used the ideas which were introduced by Ruelle (1965, 1967) for studying of thermodynamical limits in the context of the Gibbs variational principle in equilibrium classical statistical mechanics to derive Sanov's theorem as well as Banach space case of Cramér's theorem.

Donsker and Varadhan (1983) were the first to formulate and proved the large deviation principle for the distribution of the empirical process (i.e. the empirical measure of the whole process) in case the underlying process is Markovian. The solution was given in terms of a mean relative entropy as the rate function. The mean relative entropy has another interesting characterization. It was found to be closely related to the Kolmogorov–Sinai invariant of the corresponding dynamical system (see Takahashi (1982) who gives a formulation of the Gibbs variational principle in an abstract manner which unifies the Donsker and Varadhan theory for Markov chains, the equilibrium classical statistical mechanics of lattice systems and the theories for symbolic dynamics and Anosov diffeomorphisms).

There are many literatures concerning the large deviation principle for the occupation time functional and the empirical measure. The reader is referred to Varadhan (1984) and Deuschel and Stroock (1989) for a detailed discussion and for complete references. Lanford (1973) and Ellis (1985) look at large deviations from the point of view of statistical mechanics.

The first large deviation result for distribution on a function space was obtained by Schilder (1966) for a family of Wiener processes $(\varepsilon w_t)_{t\leq T}$, $\varepsilon > 0$. If we denote $\rho(\varphi, \psi)$ the uniform metric on the space of continuous function $C[0,T]$, it is reasonable to consider the logarithmic asymptotics (as $\varepsilon \to 0$) of the probabilities $\mathsf{P}(\rho(\varepsilon w, \varphi) \leq \delta)$, $\varphi \in C[0,T]$, $\delta > 0$, for processes $(\varepsilon w_t)_{t\leq T}$ to stay in tube neighbourhoods of different curves. The rate function for this problem was found to be

$$\frac{1}{2}\int_0^T |\dot{\varphi}(t)|^2 \, dt,$$

in a good accordance with a heuristic representation of Wiener's measure

$$W(d\varphi) = c \exp\left(-\frac{1}{2}\int_0^T |\dot{\varphi}(t)|^2 \, dt\right) d\varphi.$$

Varadhan (1966) deals with diffusion processes without drift and with a small diffusion term. Borovkov (1967) studied random processes with independent stationary increments. Wentzell and Freidlin (1970, 1972, 1984) were the first to undergo a systematic study of dynamical systems with small random perturbations via large deviations. Particularly, they proved the large deviation principle and mean exit time theorem for infinitely divisible random processes.

Freidlin (1978) formulated and proved the large deviation principle for the family of processes $X^\varepsilon = (X_t^\varepsilon)_{t\leq T}$ with

$$\dot{X}_t^\varepsilon = b(X_t^\varepsilon, \xi_{t/\varepsilon}),$$

where the 'fast' process $\xi_{t/\varepsilon}$ has good ergodic properties. We will refer to this processes as the processes with averaging. The main hypothesis under which he worked is that there exists a function $G(x, \alpha)$, $x \in R^r$, $\alpha \in R^r$, such that for step-functions φ_s, α_s, $s \in [0, T]$,

$$\lim_{\varepsilon\to 0} \varepsilon \ln \mathsf{E} \exp\left[\varepsilon^{-1}\int_0^T (\alpha_s, b(\varphi_s, \xi_{s/\varepsilon}))\, ds\right] = \int_0^T G(\varphi_s, \alpha_s)\, ds,$$

(here (x, y) is the inner product).

The use of the assumption like mentioned above was introduced by Gärtner (1977) in an attempt to replace the independence with some ergodicity property in (finite dimensional) Cramér's and Sanov's theorems. The problem of the existence of such a limit is closely related to the problem of thermodynamical limits in statistical mechanics (see Ruelle (1965, 1967), Lanford (1973) and Ellis (1985)).

In the context of Cramér-type problems the hypotheses takes the following form: there exists a function $G(\alpha)$, $\alpha \in R^r$, such that

$$\lim_{n\to\infty} n^{-1} \ln \mathsf{E} \exp\left[\left(\alpha, \sum_{i=1}^{n} \xi_i\right)\right] = G(\alpha).$$

It turns out that under the Gärtner assumption the Cramér-type result holds with the rate function defined as the Legendre transform of $G(\alpha)$.

Nevertheless, it is not an easy task to check the Freidlin and Gärtner hypotheses in a general case and the second part of our book (Chapters 8–11) is mainly devoted to this problem. Freidlin's approach, which is based on a Frobenius-type theorems for positive operators, seems to be not very popular in large deviations. Perhaps, this situation is caused by restrictive compactness assumptions inherent in such theorems. However it turns out and we show it in the second part of the book that Freidlin's method also works well in non-compact situations.

It could be useful to note that normalized partial sums one may represent with the help of the occupation time functional μ_n as follows:

$$n^{-1} \sum_{k=1}^{n} \xi_k = \int y \mu_n(\mathrm{d}y).$$

Therefore having got the occupation time large deviation results for Markov and stationary processes (Donsker and Varadhan (1975), Olla (1987), Orey and Pelikan (1988) and Ney and Nummelin (1987)), one may attempt to obtain extensions of Cramér's theorem via this representation. Our results show that the Gärtner and Freidlin approach allows us to prove the large deviation principle in a unified form both for Markov and stationary processes under similar (not identical in general) conditions as in cited works. This approach enables us to prove the large deviation principle for non-compact averaging processes (Theorem 11.4), which is in fact one of our main goals.

In the present book we mainly deal with extensions of Cramér's theorem for dependent cases and with large deviations for discrete time processes with averaging. Particularly we consider the family $X^n = (X_t^n)$, $t = 1, 2, \ldots$, with

$$X_t^n - X_{t-1}^n = n^{-1} b(X_{t-1}^n, \xi_t),$$

and an ergodic sequence (ξ_t), which may depend on X_t^n. As far as we know there is only one book (Wentzell and Freidlin (1984)) in which the last problem (in continuous time context) is systematically studied (see also Azencott (1980) for the treatment of large deviations for randomly perturbed dynamical systems). The main examples behind this problem for us were recurrent algorithms of identification and adaptive control (see Ljung (1977) and Kushner and Clark (1978)).

The book is organized as follows. Chapter 1 may be viewed as introduction to large deviations. We formulate and prove some basic results and discuss the roots and application of the theory. Section 1.1 contains the general idea and definition of the large deviation principle, and formulates and proves Cramér's theorem and discusses the problem how to replace the independence by some other conditions. In this way we formulate the Gärtner theorem. In Section 1.2 we present short discussion of the large deviation problem for the occupation time functional. To show as soon as possible how large deviations work, armed with this result, in Section 1.3 we apply them to some simple problems of statistical mechanics. The other goal of this consideration was to show the deep relations between large deviations and certain foundational questions in statistical mechanics (see Lanford (1973) and Ellis (1985) for detailed discussion). In Section 1.4 we present a general framework for studying large deviations. In particular, we emphasize the important role of the exponential tightness in proving the large deviation principle for distributions on function space. We next move to the problem of large deviation of function space and applications to statistics and adaptive control and identification in Sections 1.5 and 1.6.

In Chapter 2 we prove the large deviation principle for the averaging processes with additive white noise for the case when coefficients may depend on the whole past, i.e., for the following recurrent scheme:

$$X_t^n - X_{t-1}^n = n^{-1}[b_t(X^{n,t-1}) + \sigma_t(X^{n,t-1})\xi_t],$$

where $X^{n,t-1} = (X_0^n, X_1^n, \dots, X_{t-1}^n)$ and $(\xi_t)_{t\geq 1}$ is an i.i.d sequence.

The contents of Chapter 3 may be viewed as an example of the large deviation result for a stationary processes. More precisely we prove the large deviation principle for the above scheme for the case when $(\xi_t)_{t\geq 1}$ is a stationary sequence which admits the Wold-type decomposition. It turns out that in this specific case it is not necessary to use Freidlin's assumption to obtain the large deviation principle.

The proofs in Chapters 2 and 3 are based on some ideas and concepts of the martingale approach, in particular, on the concepts of stochastic cumulant. In Chapter 4 we use this notation to formulate and prove an extension of Cramér's theorem.

Chapter 5 contains some auxiliary but having independent value results concerning some bounds of a mixing rate for a class of Markov processes which we make use of next to prove the averaging principle (i.e. the law of large numbers) in Chapter 6 and the normal deviations in Chapter 7 for the processes with averaging.

Chapter 8 is devoted to an extension of Cramér's theorem for Markov processes based on the Gärtner and Freidlin approach. Chapter 9 contains a similar exposition for non-Markovian stationary processes.

In Chapter 10 following Gärtner approach we obtain the large deviation principle for empirical measures for Markov and stationary processes. Since this problem does not belong to the mainstream of the book, the exposition of the chapter is not detailed.

Chapter 11 deals with the large deviation principle for general processes with averaging both in compact and non-compact cases. The exposition is based on the results of Chapter 8.

Chapters 1 through 4 and 6 are mainly written by O. V. Gulinsky, Chapters 5 and 7 through 11 by A. Yu. Veretennikov.

The authors are indebted to Prof. R. Sh. Liptser for useful discussions.

Chapter 1

Introduction to large deviations

1.1. CRAMÉR-TYPE RESULTS

Let $\xi = (\xi_k)$, $k \geq 1$, be a sequence of independent and identically distributed (i.i.d.) random variables with a common distribution μ that are defined on a probability space (Ω, F, P) and takes value in R. Let

$$S_n = \sum_{k=1}^{n} \xi_k,$$

$n = 1, 2, \ldots$, and assume that $\int_R |u| \mu(du)$ is finite. The strong law of large numbers states that

$$S_n/n \longrightarrow m,$$

P-a.s., where m is the mean $\int_R u\mu(du)$. If F_n denotes the distribution of S_n/n on R, then as $n \to \infty$, (F_n), $n \geq 1$, converges weakly to the unit point measure concentrated at some point m in R; i.e., $F_n \Rightarrow \delta_m$ as $n \to \infty$. Thus if A is any Borel set in R whose closure $\bar{A}$ does not contain m, then $F_n(A) \to 0$; and so we can say that 'events' $A \subset R$ lying outside of a neighbourhood of m are 'deviant'. It is often an important problem to determine how 'deviant' a particular event is. That is, given A such that $m \notin \bar{A}$ one wants to know the rate at which $F_n(A)$ is tending to 0.

Following Ellis (1985) we will refer to the described problem as *Level-1* for sums of random variables.

The theory of large deviations is concerned with exponential decay of a family of probability measures $F_n(A)$. In the problems we will consider, $F_n(A)$ converges to 0 exponentially fast as $n \to \infty$ with an exponential rate depending on the set A.

The situation is abstracted in the following definition (Varadhan (1966)).

Definition 1.1. Let X be a complete separable metric space with a metric ρ, $\mathcal{B}(X)$ the Borel σ-field of X and (F_n), $n = 1, 2, \ldots$, a sequence of probability measures on $\mathcal{B}(X)$. The sequence (F_n) is said to have large deviation property if there exists a sequence of positive real member (a_n) which tend to ∞ and a function $I(\cdot)$ which map X into $[0, \infty]$ such that the following hold:

(0) for each real number $b > 0$ the set

$$\Phi(b) = (x \in X\colon\ I(x) \leq b)$$

is a compact set in X;

(I) for each open set G in X

$$\liminf_{n\to\infty} a_n^{-1} \ln F_n(G) \geq -\inf_{x\in G} I(x);$$

(II) for each closed set K in X

$$\limsup_{n\to\infty} a_n^{-1} \ln F_n(K) \leq -\inf_{x\in K} I(x),$$

$I(x)$ is called *a rate function* or *an entropy function* of F_n.

In the equivalent Wentzell and Freidlin's approach to large deviations (see Freidlin and Wentzell (1984)) conditions (I′) and (II′) are used instead of (I) and (II):

(I′) for all $\delta > 0$ and $x \in X$

$$\liminf_n a_n^{-1} \ln F_n(y\colon\ \rho(x,y) < \delta) \geq -I(x);$$

(II′) for all $\delta > 0$ and $b > 0$

$$\limsup_n a_n^{-1} \ln F_n(y\colon\ \rho(y,\Phi(b)) \geq \delta) \leq -b.$$

The reader is referred to Section 1.4 for detailed discussion of the different approaches to large deviations.

Given a Borel set A denote its closure as $\mathrm{cl}\, A$ and its interior as $\mathrm{int}\, A$. If $I(\mathrm{cl}\, A) = I(\mathrm{int}\, A)$ then the large deviation principle implies that

$$\lim a_n^{-1} \ln F_n(A) = -\inf_{x\in A} I(x).$$

As a consequence of Definition 1.1 we have the following important results, proved in Varadhan (1966).

Theorem 1.1. Assume that the sequence $(F_n)_{n\geq 1}$ of probability measures on $(X, \mathcal{B}(X))$ has the large deviation property with rate function I. Let V be a continuous function from X to R and

$$\lim_{c\to\infty} \limsup_{n\to\infty} a_n^{-1} \ln \int_{(V>c)} \exp(a_n V(x))\, F_n(\mathrm{d}x) = -\infty.$$

Then $\sup_{x\in X} [V(x) - I(x)]$ is finite and

$$\limsup_{n\to\infty} a_n^{-1} \ln \int_X \exp(a_n V(x))\, F_n(\mathrm{d}x) = \sup_{x\in X} [V(x) - I(x)].$$

We will give only heuristic motivation of the Theorem 1.1. The Definition 1.1 may be expressed by the heuristic formula

$$F_n(\mathrm{d}x) \approx \exp(-a_n I(x))\, \mathrm{d}x.$$

Hence one may write

$$\lim_{n\to\infty} a_n^{-1} \ln \int_X \exp(a_n V(x))\, F_n(\mathrm{d}x) \approx \lim_{n\to\infty} a_n^{-1} \int_X \exp[a_n(V(x) - I(x))]\, \mathrm{d}x,$$

and, by analogy with Laplace's method, one is led to the statement of Theorem 1.1.

1.1.1. The classical Cramér theorem

Let now assume that *the moment generating function*

$$\mathsf{E}\exp(\lambda\xi_1) = \int_R \exp(\lambda x)\, \mathrm{d}\mu(x) \tag{1.1}$$

is finite for all $\lambda \in R$ (Cramér's renowned condition) and introduce *the logarithmic moment generating function* of random variable ξ_1

$$G_\mu(\lambda) = \ln \int_R \exp(\lambda x)\, \mathrm{d}\mu(x) = \ln \mathsf{E}\exp(\lambda\xi_1) \qquad \text{for } \lambda \in R. \tag{1.2}$$

Note that $G_\mu(\lambda)$ is the function from $R \to (-\infty, \infty]$ with the properties:

(a) $G_\mu(\lambda)$ is convex.

Indeed, by Hölder's inequality

$$\begin{aligned} G_\mu(c\lambda_1 + (1-c)\lambda_2) &= \ln \int_R \exp(c\lambda_1 x)\, \exp((1-c)\lambda_2 x)\, \mathrm{d}\mu(x) \\ &\leq \ln \left[\left(\int_R \exp(\lambda_1 x)\, \mathrm{d}\mu(x) \right)^c \left(\int_R \exp(\lambda_2 x)\, \mathrm{d}\mu(x) \right)^{(1-c)} \right] \\ &= cG_\mu(\lambda_1) + (1-c)\, G_\mu(\lambda_2). \end{aligned}$$

(b) Under Cramér's condition $G_\mu(\lambda)$ is a smooth function.

Since $G_\mu(\lambda)$ is convex we can define the function $L(x)$ as the Legendre transform of $G_\mu(\lambda)$:

$$L(x) = \sup_\lambda\, [\lambda x - G_\mu(\lambda)]. \tag{1.3}$$

It is not hard to motivate this definition. For proving upper bound (see Definition 1.1) it is natural to apply the exponential Chebyshov inequality. Consider the set $(S_n/n \geq x)$, $x \geq m$. Then, for any $\lambda > 0$,

$$\mathsf{P}(S_n \geq nx) \leq \exp(-\lambda x)\, \mathsf{E}\exp(\lambda S_n) = \exp(-\lambda xn + \ln \mathsf{E}\exp(\lambda S_n)).$$

Since (ξ_k) is a sequence of i.i.d. random variables, we have

$$\begin{aligned} G_n(\lambda) &= \ln \mathsf{E}\exp(\lambda S_n) = \ln \prod_{k=1}^n \mathsf{E}\exp(\lambda\xi_k) \\ &= \sum_{k=1}^n \ln \mathsf{E}\exp(\lambda\xi_k) = nG_\mu(\lambda). \end{aligned} \tag{1.4}$$

Remark 1.1. Later on the logarithmic moment generating function $G_n(\lambda)$ of S_n will play an important role. Unfortunately, (1.4) holds only for i.i.d. random variables. We will see below that to replace the independence with some other property and still end up with a Cramér-type result we have to impose strong ergodicity conditions.

Taking into account (1.4) we obtain

$$\mathsf{P}(S_n \geq nx) \leq \exp\left(-n(\lambda x - G_\mu(\lambda))\right)$$

and by optimizing the last inequality, we arrive at the following estimation:

$$\mathsf{P}(S_n \geq nx) \leq \exp\left(-n \sup_{\lambda \geq 0} \left[\lambda x - G_\mu(\lambda)\right]\right). \tag{1.5}$$

It remains to show that

$$\sup_{\lambda \geq 0} \left[\lambda x - G_\mu(\lambda)\right] = \sup_{\lambda \in R} \left[\lambda x - G_\mu(\lambda)\right] = L(x)$$

and extend (1.5) to cover all closed set K. To this end we present the following Lemma which contains some useful properties of $L(x)$.

Lemma 1.1. (i) $L(x)$ is a continuous and convex function;

(ii) $L(m) = 0$, $m = \mathsf{E}\xi_1$, and $L(x)$ is non-decreasing on $[m, \infty)$ and non-increasing on $(-\infty, m]$;

(iii) for $x \geq m$,

$$\sup_{\lambda \geq 0} \left[\lambda x - G_\mu(\lambda)\right] = L(x),$$

and for $x \leq m$

$$\sup_{\lambda \leq 0} \left[\lambda x - G_\mu(\lambda)\right] = L(x);$$

(iv) under Cramér's condition we have

$$\frac{L(x)}{|x|} \to \infty, \qquad \text{as } |x| \to \infty.$$

Proof. The property (i) holds by definition of $L(x)$ as the point-wise supremum of linear functions.

Since $G_\mu(0) = 0$ and $\lambda x - G_\mu(\lambda) = 0$ for $\lambda = 0$ and every $x \in R$, then

$$\sup_{\lambda \in R} \left[\lambda x - G_\mu(\lambda)\right] = L(x) \geq 0.$$

Next note that by Jensen's inequality

$$\mathsf{E}\exp(\lambda\xi_1) \geq \exp(\lambda \mathsf{E}\xi_1) = \exp(\lambda m).$$

Therefore,

$$\lambda m - \ln \mathsf{E}\exp(\lambda\xi_1) \leq 0$$

for all λ, and so $L(m) \leq 0$. Since $L(x)$ is non-negative and convex, this proves that $L(m) = 0$ and $L(x)$ is non-decreasing on $[m, \infty)$ and non-increasing on $(-\infty, m]$.

To prove (iii) we assume that $x > m$ and $\sup_{\lambda \in R}[\lambda x - G_\mu(\lambda)]$ is attained at some $\lambda^0 < 0$. Then, since $G(\lambda) \geq m\lambda$ for all λ, we have

$$L(x) = \lambda^0 x - G_\mu(\lambda^0) \leq \lambda^0(x - m) \leq 0,$$

and we come to the contradiction.

It remains only to prove (iv). To this end, note that

$$\limsup_{x \to \infty} \frac{L(x)}{x} \geq \lambda \qquad \text{for all } \lambda > 0,$$

and

$$\liminf_{x \to -\infty} \frac{L(x)}{x} \leq -\lambda \qquad \text{for all } \lambda < 0.$$

This completes the proof of lemma.

We are now ready to prove the classical Cramér theorem.

Theorem 1.2. Suppose that

$$\lim_{|\lambda| \to \infty} \frac{G_\mu(\lambda)}{|\lambda|} = \infty. \tag{1.6}$$

Under Cramér's condition the sequence of distributions (F^n) of S_n/n on R obeys the large deviation principle with the rate function

$$I(x) = L(x) = \sup_{\lambda \in R} [\lambda x - G_\mu(\lambda)].$$

Proof. Upper bound. As a consequence of part (iii) of Lemma 1.1 and (1.5) we have the following: for $x \geq m$

$$\limsup_n n^{-1} \ln \mathsf{P}(S_n \geq nx) \leq -L(x), \tag{1.7}$$

and by the similar argument, for $x \leq m$

$$\limsup_n n^{-1} \ln \mathsf{P}(S_n \leq nx) \leq -L(x). \tag{1.8}$$

Since $L(x)$ is non-decreasing on $[m, \infty)$ and non-increasing on $(-\infty, m]$, this proves the upper bound for any closed set $K \subseteq [m, \infty) \subseteq R$ or $K \subseteq (-\infty, m]$. On the other hand, if both $K \cap [m, \infty) \neq \varnothing$ and $K \cap (-\infty, m] \neq \varnothing$ let

$$x^+ = \inf (x \geq m\colon x \in K),$$

and

$$x^- = \sup (x \leq m\colon x \in K).$$

Then

$$\begin{aligned} F_n(K) &\leq \exp(-nL(x^-)) + \exp(-nL(x^+)) \\ &\leq 2\exp\left(-n \inf_{x \in K} L(x)\right), \end{aligned}$$

and so we get the upper bound in this case also.

We end the proof of upper bound with the following remarks. The first is that

$$\lim_{c\to\infty} \limsup_n n^{-1} \ln \mathsf{P}\left(\left|\frac{S_n}{n}\right| \geq c\right) = -\infty. \tag{1.9}$$

This fact follows immediately from (1.7) and (1.8) and part (iv) of Lemma 1.1.

The second remark is that

$$\limsup_{\delta\to 0} \limsup_n n^{-1} \ln \mathsf{P}\left(\left|\frac{S_n}{n} - x\right| \leq \delta\right) \leq -L(x). \tag{1.10}$$

Evidently, this fact follows from preceding discussion. Nevertheless, we prefer to prove it in a slightly different way. Define

$$z^n(\lambda) = \exp(\lambda S_n - \ln \mathsf{E}(\lambda S_n)) = \exp\left(n\left[\lambda\frac{S_n}{n} - \ln \mathsf{E}\exp(\lambda\xi_1)\right]\right). \tag{1.11}$$

It is easy to see that

$$\mathsf{E}z^n(\lambda) = 1. \tag{1.12}$$

Then

$$\begin{aligned} 1 = \mathsf{E}z^n(\lambda) &\geq \mathsf{E}\,\mathrm{I}\left(\left|\frac{S_n}{n} - x\right| \leq \delta\right) \exp\left(n\left[\lambda\frac{S_n}{n} - G_\mu(\lambda)\right]\right) \\ &\geq \mathsf{E}\,\mathrm{I}\left(\left|\frac{S_n}{n} - x\right| \leq \delta\right) \exp(n[\lambda x - |\lambda|\delta - G_\mu(\lambda)]) \\ &= \exp(nL(x) - n|\lambda|\delta)\,\mathsf{E}\,\mathrm{I}\left(\left|\frac{S_n}{n} - x\right| \leq \delta\right), \end{aligned}$$

and so

$$\limsup_{\delta\to 0} \limsup_n n^{-1} \ln \mathsf{P}\left(\left|\frac{S_n}{n} - x\right| \leq \delta\right) \leq -L(x).$$

The importance of these remarks will become clear in Section 1.4.

Lower bound. We first note that in view of Lemma 1.1 it is sufficient to prove the following:

$$\liminf_{\delta\to 0} \liminf_n n^{-1} \ln \mathsf{P}\left(\left|\frac{S_n}{n} - x\right| < \delta\right) \geq -L(x)$$

for any $x \in R$ and $\delta > 0$. From the assumption (1.6) it follows that the supremum

$$L(x) = \sup_{\lambda\in R} [\lambda x - G_\mu(\lambda)]$$

is attained at some point λ^0 for the given arbitrary x. Then

$$L(x) = \lambda^0 x - G(\lambda^0)$$

and

$$\left.\frac{\mathrm{d}G(\lambda)}{\mathrm{d}\lambda}\right|_{\lambda=\lambda^0} = \frac{\mathsf{E}(\xi_1 \mathrm{e}^{\lambda^0\xi_1})}{\mathsf{E}\mathrm{e}^{\lambda^0\xi_1}} = x.$$

Cramér's idea for proving the lower bound is to introduce a Radon–Nikodym factor in order to make originally rare events look like typical behaviour.

Let us define a new measure as

$$dQ^{n,\lambda^0} = z^n(\lambda^0)\, d\mathsf{P}.$$

In virtue of (1.12), Q^{n,λ^0} is a probability measure. Note that (E^{n,λ^0} is the expectation with respect to (w.r.t.) Q^{n,λ^0})

$$\mathsf{E}^{n,\lambda^0}\xi_1 = \mathsf{E}z^n(\lambda^0)\xi_1 = \frac{\mathsf{E}(\xi_1\, e^{\lambda^0\xi_1})}{\mathsf{E}\, e^{\lambda^0\xi_1}} = \left.\frac{dG(\lambda)}{d\lambda}\right|_{\lambda=\lambda^0} = x.$$

We next note that in the specific case of i.i.d. random variables the independence of the sequence (ξ_k) is preserved w.r.t. the new measure Q^{n,λ^0}. Thus, by the Weak Law of Large Numbers

$$\lim_n Q^{n,\lambda^0}\left(\left|\frac{S_n}{n} - x\right| < \delta\right) = 1. \tag{1.13}$$

Armed with this results we are now ready to prove the lower bound. To this end note that the definition of $z^n(\lambda)$ implies that $z^n(\lambda) > 0$ (P-a.s.). Hence, the measures $Q^{n,\lambda}$ and P are equivalent and so

$$\begin{aligned}
\mathsf{P}\left(\left|\frac{S_n}{n} - n\right| < \delta\right) &= \mathsf{E}^{n,\lambda^0}(z^n(\lambda^0))^{-1}\,\mathrm{I}\left(\left|\frac{S_n}{n} - n\right| < \delta\right)\\
&= \mathsf{E}^{n,\lambda^0}\,\mathrm{I}\left(\left|\frac{S_n}{n} - n\right| < \delta\right)\exp\left(-n\left[\lambda^0\frac{S_n}{n} - G_\mu(\lambda^0)\right]\right)\\
&\geq \mathsf{E}^{n,\lambda^0}\,\mathrm{I}\left(\left|\frac{S_n}{n} - x\right| < \delta\right)\exp\left(-n\left[\lambda^0 x + |\lambda^0|\delta - G_\mu(\lambda^0)\right]\right)\\
&= \exp\left(-nL(x) - n|\lambda^0|\delta\right) Q^{n,\lambda^0}\left(\left|\frac{S_n}{n} - x\right| < \delta\right).
\end{aligned}$$

It follows now from (1.13) that

$$\liminf_{\delta\to 0}\ \liminf_n n^{-1}\ln \mathsf{P}\left(\left|\frac{S_n}{n} - x\right| < \delta\right) \geq -L(x).$$

This completes the proof of Theorem 1.2.

1.1.2. The extensions of Cramér's theorem

The proof of Theorem 1.2 is based on the fact that for i.i.d. random variables (see (1.4))

$$G_n(\lambda) = \ln \mathsf{E}\exp(\lambda S_n) = nG_\mu(\lambda).$$

What is to do if (ξ_k) is a sequence of random not necessarily independent variables?

Approach based on a free energy function

Gärtner's approach (Gärtner (1977), see also Ellis (1985)), is based on the assumption that for any λ there exists the limit

$$G(\lambda) = \lim_{n\to\infty} n^{-1}G_n(\lambda).$$

The function $G(\lambda)$ is closely related to the concept of free energy in statistical mechanics and is called *the free energy function* of the sequence (S_n) (see Ellis (1985) and Section 1.3 in this book).

It turns out that if a free energy function exists, then for the sequence (S_n) the upper large deviation bound is valid with the entropy function

$$L(x) = \sup_\lambda [\lambda x - G(\lambda)] .$$

If in addition $G(\lambda)$ is differentiable, then the lower large deviation bound is valid. Hence, if $G(\lambda)$ is differentiable for all λ, then the distributions of S_n/n have a large deviation property with entropy function $I(x) = L(x)$.

We now formulate Gärtner's theorem for the family of probability measures (μ_n) in E^r. The proof is given in Chapter 8.

Given $\alpha \in E^r$ define

$$G_n(\alpha) = \ln \int_{E^r} e^{(\alpha,x)}\mu_n(dx),$$

where $(\alpha, x) = \sum_{i=1}^r \alpha_i x_i$. Let a_n be a sequence of positive real numbers in R, which tends to ∞ as $n \to \infty$. Assume that for all α there exists the limit

$$G(\alpha) = \lim_n a_n^{-1} G_n(a_n \alpha).$$

Assume also that $G(\alpha)$ is lower semicontinuous, does not equal $-\infty$ and is finite in some neighbourhood of zero. Let

$$L(x) = \sup_\alpha [\alpha x - G(\alpha)] .$$

Theorem 1.3 (Gärtner (1977)). Under these assumptions the upper large deviation bound is valid, i.e., for all $b > 0$ and $\delta > 0$,

$$\limsup_n a_n^{-1} \ln \mu_n(y\colon\ \rho(y, \Phi(b)) \geq \delta) \leq -b,$$

where $\Phi(b) = (x \in E^r\colon\ L(x) \leq b)$ is a compact set in E^r. Assume in addition that G is sufficiently smooth. Then the lower large deviation bound is valid, i.e., for all $\delta > 0$ and $x \in E^r$

$$\liminf_n a_n^{-1} \ln \mu_n(y\colon\ \rho(x, y) < \delta) \geq -L(x).$$

What can happen if $G(\lambda)$ is not differentiable for all λ? It is possible to give examples (see, for example, Freidlin and Wentzell (1984)) such that the lower bound fails for a whole class of open set. In statistical mechanical applications, a non-differentiable free energy function corresponds to a phase transition (see Ellis (1985) and Section 1.3). Nevertheless, in Sections 8 and 9 we prove lower bounds under assumptions which, at least formally, do not require the differentiability of G. It should be noted that our conditions exclude mentioned counterexamples.

How to check the existence of free energy function? The simplest example is a finite Markov chain with values in $Y = (y_1, \ldots, y_N)$, $y_i \in R$ and positive transition matrix $P = (p_{ij})_{i,j=1}^N$ (see Chapter 8 and references herein). In this case the problem could be solved by applying the Frobenius theorem to the positive operator A defined as

$$Af(y_k) = \mathsf{E}_{y_k} f(\xi_1)\, e^{\gamma\xi_1} = \sum_{i=1}^{\infty} f(y_i)\, e^{\gamma y_i} p_{ki}.$$

Denote by $r(\gamma)$ the spectral radius of the matrix $P^\gamma = (p_{ki}^\gamma\colon\ p_{ki}^\gamma = \exp(\gamma)p_{ki})$. By the Frobenius theorem $r(\gamma)$ is the eigen-value of A with positive eigen-function. It is then an easy step to show (see Chapter 8) that the limit $G(\gamma)$ exists and is equal to $\ln r(\gamma)$.

Another example is Markov process which satisfies the local Doeblin-type condition, i.e., for any $x, y \in E^r$, $\Gamma \subset \mathcal{B}(E^r)$ and $C > 0$ the inequality

$$\mathsf{P}(\xi_1 \in \Gamma | X_0 = x) \le C\mathsf{P}(\xi_1 \in \Gamma | X_0 = y)$$

holds.

This approach may be extended to a stationary processes (see Chapter 9 for a detailed description and for complete references). We will turn back to the problem of the existence of free energy function in Section 1.2.

Approach based on a stochastic exponent

In Chapter 4 we consider a different approach to large deviations for *Level-1* sums of random not necessarily independent variables.

Let $X = (X_k)$, $-\infty < k < \infty$, be a strong stationary ergodic sequence with values in R and distribution μ, defined on (Ω, F, P). Define a filtration $\mathbf{F} = (F_k)$, $k \ge 0$, with

$$F_k = \sigma(X_i,\ -\infty < i < k).$$

Assume that X satisfies Cramér's condition, i.e.,

$$\mathsf{E}\, e^{\lambda X_1} < \infty \qquad \text{for all } \lambda \in R.$$

Our approach is based on the concepts of *the stochastic exponent*

$$\mathcal{E}_n(\lambda) = \prod_{k=1}^{n} \mathsf{E}\,(e^{\lambda X_k} | F_{k-1}),$$

and *the stochastic cummulant*

$$\ln \mathcal{E}_n(\lambda) = \sum_{k=1}^{n} \ln \mathsf{E}\,(\exp(\lambda X_k) | F_{k-1}).$$

Evidently, in the case of i.i.d. random variables

$$\ln \mathcal{E}_n(\lambda) = n \ln \mathsf{E} \exp(\lambda X_1) = n G_\mu(\lambda).$$

In the stationary ergodic case the cummulant is a strong stationary sequence and by ergodic theorem there exists a function

$$G_0(\lambda) = \mathsf{E} \ln \mathsf{E}\,(\exp(\lambda X_1) | F_0),$$

such that

$$\lim_n n^{-1} \sum_{k=1}^{n} \ln \mathsf{E}\,(\exp(\lambda X_k | F_{k-1}) = G_0(\lambda).$$

Under our conditions $G_0(\lambda)$ is a smooth convex function and we may define

$$I(x) = \sup_\lambda \left[\lambda x - G_0(\lambda)\right].$$

The problem arises to find a condition having as a consequence that $I(x)$ is a rate function for the sequence (S_n/n) with $S_n = \sum_{k=1}^n X_k$.

It turns out that $I(x)$ actually is a rate function if the following holds:

$$\lim_n n^{-1} \ln \mathsf{P}(|n^{-1} \ln \mathcal{E}_n(\lambda) - G_0(\lambda)| > 0) = -\infty. \tag{1.14}$$

The reader is referred to Chapter 4 for the proof.

Assumption (1.14) is extremely restrictive and one would like to have a condition which enable to consider a broad class of stochastic processes.

We assume that there exists a function $G(\lambda, y)$ such that

$$-\lim_n n^{-1} \ln \mathsf{E}\,(\mathcal{E}_n^{-1} | S_n = ny) = G(\lambda, y), \tag{1.15}$$

and prove that under (1.15) and some regularity conditions the sequence S_n/n obeys the large deviation principle with the rate function $I(x) = L(x)$ with

$$L(x) = \sup_\lambda \left[\lambda x - G(\lambda, x)\right].$$

The reader is referred to Chapter 4 for details.

1.2. LARGE DEVIATIONS ON THE SPACE OF PROBABILITY MEASURES

Level-2. The empirical measure

Sanov's theorem. Let (ξ_k), $k \geq 1$, be a sequence of i.i.d. variables, defined in Section 1.1. Let $\mathsf{M}_1(R)$ denotes the space of probability measures on the Borel subsets of R. For a Borel subset $A \subset \mathcal{B}(R)$ define the random measure

$$\mu_n(A, \omega) = n^{-1} \sum_{k=1}^{n} \mathrm{I}\,(\xi_k \in A). \tag{1.16}$$

For $A \subset \mathcal{B}(R)$ *the empirical measure* $\mu_n(\cdot, \omega)$ gives the relative frequency with which the sequence (ξ_k), $k \geq 1$ lies in A. For each ω, $\mu_n(\cdot, \omega)$ is an element of $\mathsf{M}_1(R)$. The mean of $\mu_n(\cdot, \omega)$ is the partial sum S_n/n. The strong law of large numbers implies that

$$\mu_n(\cdot, \omega) \Rightarrow \mu, \qquad \mathsf{P}\text{-a.s.}$$

Let Q_n be the distribution of $\mu_n(\cdot, \omega)$ on $\mathsf{M}_1(R)$, then

$$Q_n \Rightarrow \delta_\mu, \qquad \text{as } n \to \infty.$$

Let $\mathsf{M}_1(R)$ be supplied by a metric ρ. The large deviation problem in the present context is to find the probability (as $n \to \infty$) that the measure $(\mu_n(\cdot, \omega))$ lies close to a preassigned measure $\nu \in \mathsf{M}_1(R)$, that is, a logarithmic asymptotic of the probability of events

$$(\rho\,(\mu_n\,(\cdot,\omega),\,\nu) < \delta).$$

The large deviation property for this problem was first obtain by Sanov (1957) and then was extended by many authors (see Chapter 10 for references)

In the case that the i.i.d random variables (ξ_k), $k \geq 1$, have a finite state space $\Gamma = (x_1, \ldots, x_r)$ $(x_1, \ldots, x_r$ are r distinct real numbers), the large deviation property may be easily obtained as a consequence of Theorem 1.3 (in fact, Cramér's theorem). In this case, the distribution μ of ξ_1 is the vector $(\mu^{(1)}, \ldots, \mu^{(r)})$ with $\mu^{(i)} = \mathsf{P}(\xi_1 = x_i) > 0$. The space of probability measures $\mathsf{M}_1(\Gamma)$ is homeomorphic to compact convex subset $\mathcal{M}$ of E^r consisting of all vectors $\nu = (\nu_1, \ldots, \nu_r)$ with $\nu_i \geq 0$ and $\sum_{i=1}^r \nu_i = 1$. The empirical measure $\mu_n(\cdot,\omega)$ has the form

$$\left(n^{-1}\sum_{i=1}^{n} \mathrm{I}\,(\xi_i = x_1), \ldots, n^{-1}\sum_{i=1}^{n} \mathrm{I}\,(\xi_i = x_r)\right).$$

Let $\lambda = (\lambda_1, \ldots, \lambda_r)$ and denote

$$(n\mu_n, \lambda) = \sum_{k=1}^{r} \lambda_k \sum_{i=1}^{n} \mathrm{I}\,(\xi_i = x_k).$$

Consider

$$n^{-1}G_n(\lambda) = n^{-1} \ln \mathsf{E}\, e^{(n\mu_n,\lambda)}.$$

In the particular case of i.i.d. random variables we have:

$$\begin{aligned}
n^{-1} \ln \mathsf{E}\, e^{(n\mu_n,\lambda)} &= n^{-1} \ln \mathsf{E} \exp\left(\sum_{k=1}^{r} \lambda_k \sum_{i=1}^{n} \mathrm{I}\,(\xi_i = x_k)\right) \\
&= n^{-1} \ln \mathsf{E} \exp\left(\sum_{i=1}^{n}\sum_{k=1}^{r} \lambda_k \mathrm{I}\,(\xi_i = x_k)\right) \\
&= n^{-1} \sum_{i=1}^{n} \ln \mathsf{E} \exp\left(\sum_{k=1}^{r} \lambda_k \mathrm{I}\,(\xi_i = x_k)\right) \\
&= \ln \mathsf{E} \exp\left(\sum_{k=1}^{r} \lambda_k \mathrm{I}\,(\xi_1 = x_k)\right) \\
&= \ln \sum_{k=1}^{r} \exp(\lambda_k)\, \mu^{(k)} = G(\lambda).
\end{aligned}$$

$G(\lambda)$ is differentiable for all λ. Define

$$L(\nu) = \sup_{\lambda} \left[\sum_{k=1}^{r} \lambda_k \nu_k - \ln \sum_{k=1}^{r} \mu^{(k)} \exp(\lambda_k)\right]. \tag{1.17}$$

It is an easy step to obtain that

$$L(\nu) = \sum_{k=1}^{r} \nu_i \ln \frac{\nu_i}{\mu^{(i)}}, \tag{1.18}$$

where $0 \ln 0 = 0$. Hence, by Theorem 1.3, we get the following result.

Theorem 1.4 (Sanov (1957)). The distributions of $(\mu_n(\cdot,\omega))$ have a large deviation property on $M_1(\Gamma)$ with the entropy function

$$I(\nu) = \begin{cases} L(\nu), & \text{for } \nu \in \mathcal{M} \\ \infty, & \text{for } \nu \in E^\tau \backslash \mathcal{M}. \end{cases}$$

$I(\nu)$ is called *the relative entropy* of ν with respect to μ.

The relative entropy function had been introduced into statistics by Kullback and Leibler (1951) and plays an important role in the theory of large deviations for empirical distributions.

The basic relation between the Legendre transform of the logarithmic moment generating function $G(\lambda)$ (1.17) and the relative entropy function (1.18) is abstracted in the following definition and lemma.

Let (X,Σ) be a measurable space and let μ and ν be the probability measures on (X,Σ). Let $\mathbf{B}(\Sigma)$ be the space of bounded measurable functions f on (X,Σ).

Define

$$L(\nu) = \sup_{f \in \mathbf{B}(\Sigma)} \left[\int f(x)\,\nu(dx) - \ln \int \exp(f(x))\,\mu(dx) \right], \tag{1.19}$$

and the *relative entropy of ν with respect to μ* as

$$H(\nu|\mu) = \begin{cases} \int z(x) \ln z(x)\,\mu(dx), & \text{if } \nu \ll \mu \text{ and } z(x) = d\nu/d\mu \\ \infty, & \text{otherwise.} \end{cases}$$

Lemma 1.2. L in (1.19) is equal to $H(\cdot|\mu)$.

This lemma is implicit in the works of Ruelle (1965, 1967) and Lanford (1973) on Gibbs' variational principle. In present form it was proved by Donsker and Varadhan (1975a).

Sanov's theorem for random independent variables with values in X states that the relative entropy is a rate function for a *Level-2* large deviation problem (see, for example, Deuschel and Stroock (1989) for rigorous analysis and complete references).

The extensions of Sanov's theorem. The theory of large deviations of occupation time functional for Markov process was developed by Donsker and Varadhan (1975a, 1975b, 1977, 1983). Similar ideas and result were formulated and proved by Gärtner (1977). In our short discussion of this topic we follow Gärtner. Under assumption that a free energy function exists, the problem is solved by reducing to finite dimension case and by applying then Theorem 1.3. The main examples in Chapter 10 for which a free energy function exists are Markov and stationary processes, satisfying the Doeblin-type condition.

Let X_k, $k \geq 1$, be an ergodic sequence with values in d-dimensional Euclidean space E^d, defined on (Ω, F, P). For any Borel subset $A \in \mathcal{B}(E^d)$ define the random measure $\mu_n(A, \omega)$ as in (1.16).

Denote by $\mathbf{B}(E^d)$ the space of bounded measurable functions on E^d and define for $f \in \mathbf{B}(E^d)$ and $\mu \in \mathsf{M}(E^d)$ ($\mathsf{M}(E^d)$ is the set of all σ-additive measures on E^d)

$$\langle \mu, f \rangle = \int_{E^d} f(x)\, \mu(\mathrm{d}x).$$

Define the metric ρ on $\mathsf{M}_1(E^d)$ by

$$\rho(\mu, \nu) = \sup\, (|\langle \mu, f \rangle - \langle \nu, f \rangle| \colon f \in \mathbf{B}(E^d),\ \|f\|_{\mathbf{B}} \leq 1,\ \|f\|_{\mathrm{Lip}} \leq 1),$$

where $\mathfrak{M} = (f \in \mathbf{B}(E^d),\ \|f\|_{\mathbf{B}} \leq 1,\ \|f\|_{\mathrm{Lip}} \leq 1)$ is the class of bounded by 1 functions which satisfy the Lipschitz condition with 1 as the constant.

If μ_∞ is an invariant measure, then by the Birkhoff and Khinchin ergodic theorem, $\mu_n \Rightarrow \mu_\infty$ as $n \to \infty$.

Assume that the following condition holds.

(0) For any $\delta > 0$ there exists a finite system $\mathcal{U}$ and a finite system $\mathcal{V}$ of Borel functions on E^d with absolute value bounded by 1, such that $|\langle \mu_\infty, w \rangle| \leq \delta$, for $w \in \mathcal{V}$. In addition, for any $f \in \mathfrak{M}$ there exists $v \in \mathcal{U}$ and $w \in \mathcal{V}$ such that $|f - v| \leq w$.

Example. Let E be a compact metric space. Consider $\mathsf{M}_1(E)$ with the metric ρ defined above. Then one may choose δ-net as the finite system $\mathcal{U}$. In this case $\mathcal{V}$ consists on the single constant $\delta > 0$.

This property enables us to reduce the problem to the large deviation problem in on finite dimensional space and then apply Theorem 1.3 to complete the proof.

Gärtner's approach to the large deviation for empirical measure is based on the assumption that for any $f \in \mathbf{B}(E^d)$ there exists the limit

$$G(f) = \lim_{n \to \infty} n^{-1} \ln \mathsf{E} \exp (n \langle \mu_n, f \rangle). \tag{1.20}$$

Assume also that:

(A.1) for any $f, g \in \mathbf{B}(E^d)$ the function $h(\gamma) = G(f + \gamma g)$ is differentiable for all $\gamma \in R$;

(A.2) if a sequence of measurable functions (f_n) converges to f ($f \in \mathbf{B}(E^d)$) w.r.t. the measure μ_∞, then $G(f_n) \to G(f)$.

For $\mu \in \mathsf{M}(E^d)$ define the Legendre transformation of $G(f)$:

$$S(\mu) = \sup_f\, [\langle \mu, f \rangle - G(f)]\,, \qquad \mu \in \mathsf{M}(E^d).$$

Under conditions (0), (A.1) and (A.2) the functional $S(\mu)$ has the following properties (see Gärtner (1977) for the proof):

(B.1) $S(\mu) < \infty \Rightarrow \mu \geq 0$ and $\mu(E^d) = 1$;

(B.2) for any $s \geq 0$ the set $(\mu\colon\ S(\mu) \leq s)$ consists on the measures that are absolutely continuous w.r.t. μ_∞, and $d\mu/d\mu_\infty$ are uniformly bounded on this set;

(B.3) the functional S is lower semicontinuous in topology, induced by metric ρ, and the set $\Phi(s) = (\mu\colon\ S(\mu) \leq s)$, $s < \infty$, is compact in this topology;

(B.4) let $\mathcal{U}$ be some finite system of functions from $\mathbf{B}(E^d)$, define

$$\rho_{\mathcal{U}}(\mu, \nu) = \sup_{v \in \mathcal{U}} |\langle \mu, v\rangle - \langle v, \nu\rangle|,$$

$$S_{\mathcal{U}}(\mu) = \sup_{f \in \mathcal{L}(\mathcal{U})} (\langle \mu, f\rangle - G(f))$$

(here $\mathcal{L}(\mathcal{U})$ is a linear span of functions from $\mathcal{U}$), and

$$\Phi_{\mathcal{U}}(s) = (\mu \in \mathsf{M}(E^d)\colon\ S_{\mathcal{U}}(\mu) \leq s), \qquad s \geq 0;$$

then for any $\delta > 0$, $s \geq 0$

$$\inf\,(S\,(\nu)\colon\ \rho_{\mathcal{U}}(\nu, \mu) < \delta) \leq S_{\mathcal{U}}(\mu),$$

and

$$\Phi_{\mathcal{U}}(s) \subseteq (\mu \in \mathsf{M}(E^d)\colon\ \rho_{\mathcal{U}}(\mu, \Phi(s)) < \delta).$$

Theorem 1.5 (Gärtner (1977)). Let conditions (0), (1.20), (A.1) and (A.2) hold. Then $S(\mu)$ is the rate function in the large deviation problem for the family of random measures (μ_n) in metric space $(\mathsf{M}_1(E^d), \rho)$, i.e., for any $\mu \in \mathsf{M}_1(E^d)$ and $s > 0$

$$\liminf_{\delta \to 0}\ \liminf_{n \to \infty} n^{-1} \ln \mathsf{P}\,(\rho\,(\mu_n, \mu) < \delta) \geq -S(\mu),$$

$$\limsup_{\delta \to 0}\ \limsup_{n \to \infty} n^{-1} \ln \mathsf{P}\,(\rho\,(\mu_n, \Phi(s)) \geq \delta) \leq -s.$$

We now sketch the proof of Theorem 1.5. The reader is referred to Gärtner (1977) (see also Freidlin and Wentzell (1984)) for details.

Suppose that $S(\mu) < \infty$. By condition (0), for any $\delta > 0$ the following estimation is valid:

$$\rho\,(\mu_n, \mu) \leq \rho_{\mathcal{U}}(\mu_n, \mu) + \max_{w \in \mathcal{V}} \langle \mu_n, w\rangle + \max_{w \in \mathcal{V}} \langle \mu, w\rangle\,.$$

Since by the property (B.2) μ is absolutely continuous w.r.t. μ_∞, we have for small δ that

$$\max_{w \in \mathcal{V}} (\langle \mu, w\rangle) < \delta/4.$$

Thus,

$$\mathsf{P}\,(\rho\,(\mu_n, \mu) < \delta) \geq \mathsf{P}\,(\rho_{\mathcal{U}}(\mu_n, \mu) < \delta/2) - \mathsf{P}\left(\max_{w \in \mathcal{V}} (\langle \mu_n, w\rangle) \geq \delta/4\right).$$

By applying Theorem 1.3 to the family of finite dimension vectors $(\langle\mu_n, v\rangle)_{v\in\mathcal{U}}$ and taking into account that in a finite dimensional space all metrics are equivalent, we obtain

$$\liminf_{\delta\to 0}\liminf_{n\to\infty} n^{-1}\ln \mathsf{P}\left(\rho_{\mathcal{U}}(\mu_n,\mu)<\delta/2\right)\geq -S_{\mathcal{U}}(\mu)\geq -S(\mu).$$

It remains only to observe that, by exponential Chebyshov's inequality, taking into account the condition (1.20) and property (A.2), one can easily obtain that

$$\limsup_{n\to\infty} n^{-1}\ln \mathsf{P}\left(\langle\mu_n,w\rangle\geq \delta/4\right)\leq -S(\mu).$$

Turning to the upper bound, we note that, in virtue of condition (0) and property (B.2), for a small $\delta>0$ and $\mu\in\Phi(s)$, we have:

$$\max_{w\in\mathcal{V}}\left(\langle\mu,w\rangle\right)<\delta/4.$$

Then

$$\mathsf{P}\left(\rho\left(\mu_n,\mu\right)\geq\delta\right)\leq \mathsf{P}\left(\rho_{\mathcal{U}}(\mu_n,\Phi(s))\geq\delta/2\right)+\mathsf{P}\left(\max_{w\in\mathcal{V}}\left(\langle\mu_n,w\rangle\right)\geq\delta/4\right).$$

Let $v_1,\ldots,v_r\in\mathcal{U}$ be the basis of the linear span $\mathcal{L}(\mathcal{U})$. We next apply Theorem 1.3 to the family finite dimension vectors $(\langle\mu_n,v_1\rangle,\ldots,\langle\mu_n,v_r\rangle)$ to obtain (keeping in mind the property (B.4)) the following estimation:

$$\begin{aligned}\limsup_{\delta}\ \limsup_{n}\ & n^{-1}\ln\mathsf{P}(\rho_{\mathcal{U}}(\mu_n,\Phi(s))\geq\delta/2)\\ &\leq \limsup_{\delta}\ \limsup_{n}\ n^{-1}\ln\mathsf{P}(\rho_{\mathcal{U}}(\mu_n,\Phi_{\mathcal{U}}(s))\geq\delta/4)\\ &\leq -s.\end{aligned}$$

The proof is completed by the same way as in proving lower bound.

The main problem in Gärtner's approach is to check the existence of the limit (1.20) and its differentiability. The reader is referred to Chapter 10 for the examples which satisfy the Doeblin-type condition.

Cramér-type results via large deviations for empirical measure. In the preceding we have seen the way to prove Cramér-type results and large deviations for empirical measure with the help of Gärtner's theorem. It is instructive to note that *Level-1* large deviation results one can obtain as a corollary of the Sanov-type theorem.

This way is based on the following useful general fact.

Lemma 1.3. Let X be a separable metric space, f a measurable map from X into a second separable metric space X' with a metric ρ', and assume that there exists a sequence of continuous maps (f_l), $l\geq 1$, such that

$$\lim_{l\to\infty}\ \sup\ \{\rho'(f_l(x),f(x)):\ x\in X \text{ with } I(x)\leq c\}=0\quad\text{for each } c\in(0,\infty).$$

Assume that the family of probability measures $(F_n)_{n\geq 1}\subseteq \mathsf{M}_1(X)$ obeys the large deviation principle with a rate function I and has the property that

$$\lim_{l\to\infty}\limsup_{n\to\infty}\frac{1}{n}\ln F_n(x\in X:\ \rho'(f_l(x),f(x))\geq\delta)=-\infty$$

for each $\delta \in (0,\infty)$. Then the map I': $X' \to [0,\infty]$ given by

$$I'(x') = \inf\,[I(x):\ x \in X \text{ and } x' = f(x)], \qquad x' \in X',$$

governs the large deviations of $(F_n \circ f^{-1})_{n\geq 1}$.

The proof can be found, for example, in Deuschel and Stroock (1989, p. 37).

Let X_k, $k \geq 1$, be an ergodic sequence with values in E^d. Assume that a sequence of empirical measures $\mu_n(\cdot,\omega))$, $n \geq 1$, obeys the large deviation principle. Denote

$$S_n = \frac{1}{n}\sum_{k=1}^{n} b(X_k),$$

where b: $E^d \to E^d$ is a continuous map.

On can see that for each $n \geq 1$ and ω

$$\frac{1}{n}\sum_{k=1}^{n} b(X_k) = \int\limits_{E^d} b(y)\,\mu_n(dy,\omega). \tag{1.21}$$

Under some additional conditions the mean value functional satisfies the assumptions of Lemma 1.3, thus one can obtain the large deviation principle for distributions of (S_n), $n \geq 1$.

The representation (1.21) can also be used for checking the existence of the free energy function with the help of Theorem 1.1.

Level-3. Empirical process

Let (ξ_n), $n \geq 1$, be a sequence of i.i.d. random variables taking value 0 and 1 with probability 1/2 each. Given n and $\omega \in \Omega$ one can consider the n-tuples $(\xi_1,\dots,\xi_n)$ and extend it periodically on both sides into a double infinite sequence of period n, obtaining a point

$$X(n,\omega) = (\dots,\xi_n,\ \xi_1,\dots,\xi_n,\ \xi_1,\dots,\xi_n,\ \xi_1,\dots)$$

in $(0,1)^{\mathbf{Z}}$. Let T denotes the shift in the space of 0 or 1. For $\omega \in \Omega$ and A a subset of $(0,1)^{\mathbf{Z}}$, the empirical process is defined as

$$R_n(\omega, A) = n^{-1}\sum_{k=0}^{n-1} \mathrm{I}\,(T^k X(n,\omega) \in A),$$

and gives the relative frequency with which the sequence

$$X(n,\omega),\ TX(n,\omega),\dots,T^{n-1}X(n,\omega)$$

lies in A. $R_n(\omega,\cdot)$ is a random stationary measure on the space Ω. Its one-dimensional marginal is the empirical distribution of $(\xi_1,\dots\xi_n)$ and the k-dimensional marginal is the empirical distribution of k-tuples $(\xi_i,\dots,\xi_{i+k-1})$.

Denote P_0 the product-measure on Ω corresponding to independent observations. Then by the ergodic theorem $R_n(\omega,\cdot) \Rightarrow P_0$ in the sense of weak convergence almost surely w.r.t P_0. Denote F_n the distribution of R_n on the space of stationary processes on Ω. Then $F_n \Rightarrow \delta_{P_0}$ and one can ask for large deviations of F_n.

It turns out that the large deviation principle holds for F_n on the space of stationary measures with a rate function $I(\mu)$ defined for stationary measure μ by

$$I(\mu) = (\ln 2 - H(\mu)),$$

where $H(\mu)$ is the Shannon entropy of μ. This result is essentially the Shannon–Breiman–McMillan theorem in information theory. Moreover, this rate function is intimately related to the Kolmogorov–Sinai invariant of the described dynamical system.

We will not consider this problem in the rest of the book. The reader is referred to Varadhan (1984) for rigorous analysis.

1.3. APPLICATIONS TO STATISTICAL MECHANICS

The roots of large deviation theory lie in a number of areas such as statistics, information theory and statistical mechanics. Armed with Cramér-type results we are able to apply large deviation theory to simple statistical mechanical examples. In this section we consider simple examples of ferromagnetic models, adopted from Ellis (1985).

Let $\mathbf{Z}^d$ be an integer lattice and Λ be a hypercube in $\mathbf{Z}^d$ centered at the origin. To each site $j \in \Lambda$ there is assigned a variable ω_j, called a spin, which takes the value 1 (spin-up) or -1 (spin-down). The configuration space is the set Ω_Λ of all sequences $\omega = (\omega_j,\ j \in \Lambda)$, thus $\Omega_\Lambda = (1, -1)^\Lambda$. The coordinate function on Ω_Λ, defined $X_j(\omega) = \omega_j$, are called the spin random variables at the sites j.

Let β be a positive real number, which represents the inverse absolute temperature; $J(\cdot)$ a non-negative real-valued function on $\mathbf{Z}^d$ with the properties $J(k) = J(-k)$ for each $k \in \mathbf{Z}^d$ and $\sum_{k \in \mathbf{Z}^d} J(k) < \infty$; h a real number which represents an external magnetic field; and μ the probability measure $\frac{1}{2}\delta_1 + \frac{1}{2}\delta_{-1}$. The function J is called a ferromagnetic interaction.

For each $\omega \in \Omega_\Lambda$ the *Hamiltonian* or interaction energy is defined as

$$H_{\Lambda,h}(\omega) = -\frac{1}{2} \sum_{i,j \in \Lambda} J(i-j)\,\omega_i \omega_j - h \sum_{i \in \Lambda} \omega_i. \tag{1.22}$$

The ferromagnetic model is defined by the probability measure

$$P_{\Lambda,\beta,h}(\omega) = \exp\left[-\beta H_{\Lambda,h}(\omega)\right] \pi_\Lambda P_\mu(\omega) (Z(\Lambda,\beta,h))^{-1} \tag{1.23}$$

on the subsets of Ω_Λ, where $\pi_\Lambda P_\mu$ is the product measure on the subsets of Ω_Λ with identical one-dimensional marginals μ; $Z(\Lambda, \beta, h)$ is normalization:

$$\begin{aligned} Z(\Lambda,\beta,h) &= \int_{\Omega_\Lambda} \exp\left[-\beta H_{\Lambda,h}(\omega)\right] \pi_\Lambda P_\mu(d\omega) \\ &= \sum_{\omega \in \Omega_\Lambda} \exp\left[-\beta H_{\Lambda,h}(\omega)\right] (2^{|\Lambda|})^{-1} \end{aligned} \tag{1.24}$$

(note that $\pi_\Lambda P_\mu(\omega) = (2^{|\Lambda|})^{-1}$, where $|\Lambda|$ is the number of sites in Λ). The probability measure $P_{\Lambda,\beta,h}$ is called *a finite-volume Gibbs state* on Λ and the normalization $Z(\Lambda,\beta,h)$ is called a partition function. Define the *Gibbs free energy* as

$$\Psi(\Lambda,\beta,h) = -(\beta)^{-1}\ln Z(\Lambda,\beta,h). \tag{1.25}$$

Ferromagnetic models on $\mathbf{Z}$ are used to study a phase transition which is characterized by a positive spontaneous magnetization. Define *the total spin* in Λ as

$$S_\Lambda(\omega) = \sum_{j\in\Lambda} X_j(\omega), \tag{1.26}$$

and *the magnetization* as

$$M(\Lambda,\beta,h) = \int_{\Omega_\Lambda} S_\Lambda(\omega)\, P_{\Lambda,\beta,h}(\mathrm{d}\omega).$$

Note that the magnetization per site, $|\Lambda|^{-1}M(\Lambda,\beta,h)$, and the Gibbs free energy per site, $|\Lambda|^{-1}\Psi(\Lambda,\beta,h)$, are related by the formula

$$|\Lambda|^{-1}M(\Lambda,\beta,h) = -|\Lambda|^{-1}\frac{\partial\Psi(\Lambda,\beta,h)}{\partial h}. \tag{1.27}$$

The problem is to study $|\Lambda|^{-1}\Psi(\Lambda,\beta,h)$ and $|\Lambda|^{-1}M(\Lambda,\beta,h)$ in the limit as the hypercube Λ expand to fill $\mathbf{Z}^d$.

The *specific Gibbs free energy* is defined as

$$\psi(\beta,h) = \lim_{\Lambda\uparrow\mathbf{Z}^d} |\Lambda|^{-1}\Psi(\Lambda,\beta,h), \tag{1.28}$$

and *the specific magnetization* as

$$m(\beta,h) = \lim_{\Lambda\uparrow\mathbf{Z}^d} |\Lambda|^{-1}M(\Lambda,\beta,h).$$

We first consider a simple model which is ideal for doing exact calculations.

The Curie-Weiss model

In order to ease the notation we replace Λ by the set $(1,2,\ldots,n)$ and later on index all quantities by n instead of by Λ.

Define the Hamiltonian as

$$\begin{aligned} H^{CW}_{n,h}(\omega) &= -\frac{1}{2n}\sum_{i,j\in\Lambda}\omega_i\omega_j - h\sum_{i\in\Lambda}\omega_i \\ &= -n\left(\frac{1}{2}\left[n^{-1}\sum_{j=1}^{n}\omega_j\right]^2 + hn^{-1}\sum_{j=1}^{n}\omega_j\right). \end{aligned} \tag{1.29}$$

It means that the interaction for Curie–Weiss model is given by $J_\Lambda(i-j) = |\Lambda|^{-1}$, if $i,j\in\Lambda$, and $J_\Lambda(i-j) = 0$, if i or $j \notin \Lambda$. Note that in this specific case the interaction depends on the set Λ.

The Curie–Weiss model is defined by the probability

$$P^{CW}_{n,\beta,h}(\omega) = \exp\left[-\beta H^{CW}_{n,h}(\omega)\right] \pi_n P_\mu(\omega)(Z^{CW}(n,\beta,h))^{-1}.$$

Let F_n denote the distribution of the total spin

$$S_n(\omega) = \sum_{j\in\Lambda} X_j(\omega) = \sum_{j=1}^{n} \omega_j$$

w.r.t. the product measure $\pi_n P_\mu$.

The problem is to determine $\mathsf{E}^{\beta,h}_n f(S_n/n)$ as $n \to \infty$ for any continuous function f. Note that if $f(x) = x$, for $-1 \le x \le 1$, then the specific magnetization for Curie–Weiss model is given by

$$m^{CW}(\beta,h) = \lim_{n\to\infty} \mathsf{E}^{\beta,h}_n f(S_n/n).$$

By (1.29) we have

$$\begin{aligned} \mathsf{E}^{\beta,h}_n f\left(\frac{S_n}{n}\right) &= \int_{\Omega_n} f\left(\frac{S_n}{n}\right) \exp\left(-\beta H^{CW}_{n,h}(\omega)\right) \pi_n P_\mu(\mathrm{d}w)(Z^{CW}(n,\beta,h))^{-1} \\ &= \int_R f(x) \exp\left[n\left(\frac{1}{2}\beta x^2 + \beta h x\right)\right] F_n(\mathrm{d}x)(Z^{CW}(n,\beta,h))^{-1}, \quad (1.30) \end{aligned}$$

$$Z^{CW}(n,\beta,h) = \int_R \exp\left[n\left(\frac{1}{2}\beta x^2 + \beta h x\right)\right] F_n(\mathrm{d}x). \quad (1.31)$$

For real λ define *the logarithmic moment generating function* of S_n (compare with (1.4), Section 1.1)

$$\begin{aligned} G^{\beta,h}_n(\lambda) &= \ln \int_{\Omega_n} \exp\left(\lambda S_n(\omega)\right) P_{n,\beta,h}(\mathrm{d}\omega) \\ &= \ln \int_R \exp\left[n\left(\lambda x + \frac{1}{2}\beta x^2 + \beta h x\right)\right] F_n(\mathrm{d}x)(Z^{CW}(n,\beta,h))^{-1}. \end{aligned}$$

Since $\pi_n P_\mu$ is a product measure, we have

$$\begin{aligned} n^{-1} G_n(\lambda) &= n^{-1} \ln \int_{\Omega_n} \exp\left(\lambda \sum_{i=1}^{n} \omega_i\right) \pi_n P_\mu(\mathrm{d}\omega) \\ &= n^{-1} \sum_{i=1}^{n} \ln \int_R \exp(\lambda x)\, \mu(\mathrm{d}x) = G_\mu(\lambda), \end{aligned}$$

where

$$G_\mu(\lambda) = \ln \int_R \exp(\lambda x)\, \mathrm{d}\mu(x).$$

Recall that μ is probability measure $1/2\delta_{-1} + 1/2\delta_1$ on R. Then

$$G_\mu(\lambda) = \ln \cosh \lambda.$$

By Cramér's theorem the distributions (F_n) obeys the large deviation principle with entropy function

$$I(x) = \sup_\lambda (\lambda x - \ln \cosh \lambda) = \begin{cases} \frac{1+x}{2} \ln(1+x) + \frac{\ln(1-x)}{2}, & \text{for } |x| \le 1 \\ \infty, & \text{for } |x| > 1. \end{cases}$$

It means that, for large n, $F_n(dx)$ may be expressed by the following heuristic formula

$$F_n(dx) \approx \exp[-nI(x)]\, dx.$$

Define

$$I_{\beta,h}(x) = -\left(\frac{1}{2}\beta x^2 + \beta h x\right) + I(x).$$

Hence, for large n, from (1.30) and (1.31) we obtain the heuristic formula

$$\mathsf{E}_n^{\beta,h} f(S_n/n) = \frac{\int_R f(x) \exp[-nI_{\beta,h}(x)]\, dx}{\int_R \exp[-nI_{\beta,h}(x)]\, dx}.$$

By analogy with Laplace's method one may expect that the limit of $\mathsf{E}_n^{\beta,h} f(S_n/n)$ as $n \to \infty$ is determined by the points x at which the function $I_{\beta,h}(x)$ attains its infinum. This hypothesis is in fact correct. The reader is referred to Ellis (1985) for a rigorous analysis and a detailed description of the phase transition.

The Ising model

Turning back to the general model, for $\lambda \in R$ define

$$G_\mu^{\Lambda,\beta,h}(\lambda) = \ln \int_{\Omega_\Lambda} \exp(\lambda S_\Lambda(\omega))\, P_{\Lambda,\beta,h}(d\omega). \tag{1.32}$$

Since

$$\lambda S_\Lambda(\omega) - \beta H_{\Lambda,h}(\omega) = H_{\Lambda,h+\lambda\beta^{-1}}(\omega),$$

it follows from (1.23)–(1.25) that

$$\begin{aligned} G_\mu^{\Lambda,\beta,h}(\lambda) &= \ln Z(\Lambda,\beta,h+\lambda\beta^{-1}) + \ln Z(\Lambda,\beta,h) \\ &= -\beta[\Psi(\Lambda,\beta,h+\lambda\beta^{-1}) - \Psi(\Lambda,\beta,h)]. \end{aligned} \tag{1.33}$$

Define the *free energy function* of the sequence (S_Λ) with respect to the measures $(P_{\Lambda,\beta,h})$

$$\begin{aligned} G_\mu^{\beta,h}(\lambda) &= \lim_{\Lambda\uparrow \mathbf{Z}^d} |\Lambda|^{-1} G_\mu^{\Lambda,\beta,h}(\lambda) = \lim_{\Lambda\uparrow \mathbf{Z}^d} |\Lambda|^{-1} \ln \int_{\Omega_\Lambda} \exp(\lambda S_\Lambda(\omega))\, P_{\Lambda,\beta,h}(d\omega) \\ &= \lim_{\Lambda\uparrow \mathbf{Z}^d} |\Lambda|^{-1} \ln Z(\Lambda,\beta,h+\lambda\beta^{-1}) + \lim_{\Lambda\uparrow \mathbf{Z}^d} |\Lambda|^{-1} \ln Z(\Lambda,\beta,h) \end{aligned} \tag{1.34}$$

Evidently, from (1.28) and (1.33) we have that

$$G_\mu^{\beta,h}(\lambda) = -\beta[\psi(\beta,h+\lambda\beta^{-1}) - \psi(\beta,h)].$$

The interaction for the Ising model is given by $J(k) = J$, if $\|k\| = 1$, and $J(k) = 0$, if $\|k\| \neq 1$; J is a fixed positive number and $\|\cdot\|$ is the Euclidean norm on $\mathbf{Z}^d$.

We now sketch the main facts (see Ellis (1985) for details) that highlight some problems of the large deviation theory and profound relations between large deviations and statistical mechanics.

(a) $m(\beta, h)$ exists for each $\beta > 0$, $m(\beta, 0) = 0$ and $m(\beta, -h) = -m(\beta, h)$.

(b) The limit $m(\beta, +) = \lim_{h \to 0+} m(\beta, h)$ exists and is non-negative; $m(\beta, +)$ is called *the spontaneous magnetization*.

(c) There exists a positive number β_c, which may equal ∞, such that $m(\beta, +) = 0$ for $0 < \beta < \beta_c$ and $m(\beta, +) > 0$ for $\beta > \beta_c$; β_c is called *the critical inverse temperature*.

(d) The limit

$$\psi(\beta, h) = \lim_{\Lambda \uparrow \mathbf{Z}^d} (-1)|\Lambda|^{-1} \ln Z(\Lambda, \beta, h)$$

exists and for each $\beta > 0$ is a continuously differentiable function of $h \neq 0$. For $\beta > 0$ and $h \neq 0$

$$m(\beta, h) = -\frac{\partial \psi(\beta, h)}{\partial h}.$$

(e) For $0 < \beta < \beta_c$, $\psi(\beta, h)$ is differentiable at $h = 0$ and

$$m(\beta, +) = -\frac{\partial \psi(\beta, 0)}{\partial h} = 0.$$

For $\beta > \beta_c$, $\psi(\beta, h)$ is not differentiable at $h = 0$; $\partial \psi(\beta, h)/\partial h^+$, the right-hand derivative at $h = 0$, exists and

$$m(\beta, +) = -\frac{\partial \psi(\beta, 0)}{\partial h^+} > 0.$$

According to (c), there is a positive spontaneous magnetization $m(\beta, +)$ for β exceeding the critical inverse temperature β_c.

The most important facts from the large deviation theory point of view are (d) and (e). In particular, these facts show that for $\beta > 0$, $h \neq 0$ and $0 < \beta < \beta_c$, $h = 0$, the limit $\Lambda \uparrow \mathbf{Z}^d$ and differentiation with respect to h may be interchanged in (1.27).

Turning back to the free energy function (1.34), we note that from the preceding discussion it follows that the limit $G_\mu^{\beta,h}(\lambda)$ exists for $0 < \beta < \beta_c$ and any real h; $G_\mu^{\beta,h}(\lambda)$ is a differentiable function of $\lambda \in R$. By Theorem 1.3, for $0 < \beta < \beta_c$ and any real h, the $P_{\Lambda,\beta,h}$-distributions of (S_Λ/Λ) have a large deviation property with entropy function

$$I(x) = L^{\beta,h}(x) = \sup_\lambda \left[\lambda x - G_\mu^{\beta,h}(\lambda)\right].$$

1.4. BASIC LARGE DEVIATIONS CONCEPTS

In this section we discuss basic approaches to the large deviations. We first recall for convenience Varadhan's and Wentzell and Freidlin' s definitions. (We assume hereinafter for the simplicity $a_n = n$.)

Let X be a complete separable metric space with metric ρ and (F_n), $n = 1, 2, \ldots$, a sequence of a probability measures on the Borel subsets $\mathcal{B}(X)$ of X. Suppose that there exists a function $I(\cdot)$ which maps X into $[0,\infty]$ such that the following condition holds:

(0) for each real number $b > 0$ the level set

$$\Phi(b) = (x \in X\colon\ I(x) \leq b)$$

is a compact set in X.

Definition 1.2 (Varadhan (1966)). The sequence (F_n) obeys the large deviation principle with the rate function $I(\cdot)$ if the following hold:

(I) for each open set G in X

$$\liminf_{n\to\infty} n^{-1} \ln F_n(G) \geq -\inf_{x\in G} I(x),$$

(II) for each closed set K in X

$$\limsup_{n\to\infty} n^{-1} \ln F_n(K) \leq -\inf_{x\in K} I(x).$$

Definition 1.3 (Wentzell and Freidlin (1984)). The sequence (F_n) obeys the large deviation principle with the rate function $I(\cdot)$ if the following hold:

(I′) for all $\delta > 0$ and $x \in X$

$$\liminf_{n\to\infty} n^{-1} \ln F_n(y\colon\ \rho(x,y) < \delta) \geq -I(x),$$

(II′) for all $\delta > 0$ and $b > 0$

$$\limsup_{n\to\infty} n^{-1} \ln F_n(y\colon \rho(y,\Phi(b)) \geq \delta) \leq -b.$$

The following simple theorem shows that both definitions are equivalent.

Theorem 1.6 (Wentzell and Freidlin (1984)). Conditions (I) and (I′) are equivalent. The condition (II′) follows from (II). Conditions (0) and (II′) imply (II).

Proof. (I′) $\Leftrightarrow$ (I) follows immediately from the definitions. Let us show that (II) $\Rightarrow$ (II′). Since the set $K = (y\colon\ \rho(y,\Phi(b)) \geq \delta)$ is closed, we have that $I(x) > b$ for $x \in K$ and $\inf_{y\in K} I(y) \geq b$. Then, by condition (II),

$$\limsup_{n\to\infty} n^{-1} \ln F_n(K) \leq -b.$$

We next show that conditions (0) and (II′) imply (II). Pick $\gamma > 0$ and define

$$b = \inf_{y\in K} I(y) - \gamma.$$

Since K is closed, $\Phi(b)$ is a compact set and $K \cap \Phi(b) = \varnothing$, we have that $\rho(K,\Phi(b)) = \delta > 0$. Then, by condition (II′)

$$\begin{aligned}\limsup_{n\to\infty} n^{-1} \ln F_n(K) &\leq \limsup_{n\to\infty} n^{-1} \ln F_n(y\colon\ \rho(y,\Phi(b)) \geq \delta)\\ &\leq -b + \gamma = -\inf_{y\in K} I(y) + 2\gamma.\end{aligned}$$

Since γ is arbitrary we can let $\gamma \to 0$ and obtain (II).

Remark 1.2. The condition (0) is equivalent to the following conditions:

(a) $I(x)$ is a lower semicontinuous function (it means that $\Phi(b)$ is a closed set),
(b) $\Phi(b)$ is a relative compact set.

We next reproduce from Wentzell and Freidlin (1984) the following useful result.

Theorem 1.7. Let the sequence (F_n) obey the large deviation principle with the rate function $I(x)$. Then, for each $x \in X$, $I(x)$ is equal to

$$-\liminf_{\delta\downarrow 0} \liminf_{n\to\infty} n^{-1} \ln F_n(y\colon\ \rho(x,y) < \delta)$$
$$= -\limsup_{\delta\downarrow 0} \limsup_{n\to\infty} n^{-1} \ln F_n(y : \rho(x,y) < \delta) \quad (= J(x)). \quad (1.35)$$

Proof. By conditions (I) or (I′) it follows immediately that

$$\liminf_{\delta\downarrow 0} \liminf_{n\to\infty} n^{-1} \ln F_n(y\colon\ \rho(x,y) < \delta) \geq -I(x).$$

We show that (0) and (II) or (II′) imply

$$\limsup_{\delta\downarrow 0} \limsup_{n\to\infty} n^{-1} \ln F_n(y\colon \rho(x,y) < \delta) \leq -I(x).$$

To this end, pick $\gamma > 0$ and note that the compact set $\Phi(I(x)-\gamma)$ does not contain x. Choose $\delta > 0$ such that $\delta < 1/2\rho(\Phi(I(x)-\gamma), x)$. Then, by condition (II′)

$$\limsup_{\delta\downarrow 0} \limsup_{n\to\infty} n^{-1} \ln F_n(y\colon\ \rho(x,y) < \delta)$$
$$\leq \limsup_{\delta\downarrow 0} \limsup_{n\to\infty} n^{-1} \ln F_n(y\colon\ \rho(\Phi(I(x)-\gamma), y) \geq \delta) \leq I(x) - 2\gamma.$$

Since γ is arbitrary by letting $\gamma \to 0$ we obtain the desired result.

We will refer to (1.35) as *the local large deviation principle.*

It would be convenient to reduce the analysis of large deviations to the local large deviations principle. However, generally speaking the local large deviation principle does not imply the full one.

Example (Wentzell and Freidlin (1984)). Let $X \equiv R$, $\rho(x,y) = |x-y|$ and

$$F_n(x) = c\mathcal{N}(0, n^{-1}) + c\mathcal{N}(0, \sigma(n))$$

with $\sigma(n) = \exp(\mathrm{e}^{n^2})$. It is easy to see that

$$\lim_{\delta\downarrow 0} \lim_{n\to\infty} n^{-1} \ln F_n(y\colon\ \rho(x,y) < \delta) = -\frac{x^2}{2}$$

for every $x \in R$. The function $I(x) = x^2/2$, evidently, satisfies the condition (0). However, conditions (II) and (II′) do not hold.

Nevertheless, it is possible to show that under the exponential tightness condition the local large deviation principle does imply the full one.

Definition 1.4. The sequence of measures (F_n), $n \geq 1$, is said to be *exponentially tight* if, for each $L > 0$, there is a compact set K_L in X such that

$$\limsup_n n^{-1} \ln F_n(K_L^c) \leq -L.$$

The approach for proving the large deviation principle, based on the concept of exponential tightness and the weak large deviation principle, was successively used by Deuschel and Strook (1989).

Definition 1.5. The sequence (F_n) is said to have *the weak large deviation property* if there exists a lower semicontinuous function $I(\cdot)$ which maps X into $[0, \infty]$ such that the condition (I) holds and for each *compact* set K in X

$$\limsup_{n\to\infty} n^{-1} \ln F_n(K) \leq - \inf_{x\in K} I(x).$$

The passage from a weak to a full large deviation principle is accomplished by an application of the following result.

Theorem 1.8. Let the sequence (F_n), $n \geq 1$, satisfy the weak large deviation principle. If the sequence (F_n) is exponentially tight, then it also obeys the full large deviation principle.

Proof. First note that

$$\inf_{x\in K_L^c} I(x) \geq - \liminf n^{-1} \ln F_n(K_L^c) \geq L$$

and so

$$\Phi(L) = (x\colon\ I(x) \leq L) \subseteq K_{L+1}.$$

Since $I(x)$ is lower semicontinuous, this proves that the condition (0) holds.

Let G be a closed subset X and set

$$G_L = G \cap K_L$$

for $L \geq 0$. Then

$$F_n(G) \leq F_n(G_L) + F_n(K_L^c)$$

and so

$$\lim_n n^{-1} \ln F_n(G) \leq - \left(\inf_{x\in G_L} I(x) \wedge L\right) \leq - \left(\inf_{x\in G} I(x) \wedge L\right)$$

for every $L > 0$. Thus, by letting $L \to \infty$, we get the required result.

Pukhalskii (1991) proved for large deviations an analog of Prokhorov theorem on the equivalence of weak relative compactness and tightness for a family of probability measures. In our presentation in the rest of this section we follow Pukhalskii (1991) and Liptser and Pukhalskii (1992).

We need the following definition.

Definition 1.6. The sequence (F_n), $n \geq 1$, is said to have *the partial large deviation principle* if any subsequence (n') of (n) contains a further subsequence (n'') such that $(F_{n''})$ obeys the large deviation principle with some rate function.

Theorem 1.9. A sequence (F_n), $n \geq 1$, is *exponentially tight* if and only if it satisfies *the partial large deviation principle.*

The reader is referred to Pukhalskii (1991) for the proof.

Later on we assume that $X = D[0,1]$, since mostly this space is required in the rest of the book. Recall that $D[0,1]$ is the Skorokhod space of all right continuous having left hand limits real-valued functions $X = (X_t)_{0 \leq t \leq 1}$.

Let $X^n = (X_t^n)_{t \leq 1}$, $n \geq 1$, be a sequence of random processes with paths in $D[0,1]$, defined on a probability space (Ω, F, P).

Definition 1.7. A sequence $X^n = (X_t^n)_{t \leq 1}$, $n \geq 1$, of random processes with paths in $D[0,1]$ is said to satisfy *the local large deviation principle in* $D[0,1]$ if for any $\varphi = (\varphi_t)_{t \leq 1} \in D[0,1]$ the following equality holds:

$$-\liminf_{\delta \downarrow 0} \liminf_{n \to \infty} n^{-1} \ln \mathsf{P}\,(\rho\,(X^n, \varphi) \leq \delta)$$
$$= -\limsup_{\delta \downarrow 0} \limsup_{n \to \infty} n^{-1} \ln \mathsf{P}\,(\rho\,(X^n, \varphi) \leq \delta) \quad (= J(\varphi)).$$

The following theorem establishes that under exponential tightness the local large deviation principle implies the full one.

Theorem 1.10. If a sequence $X^n = (X_t^n)_{t \leq 1}$, $n \geq 1$, of random processes with paths in $D[0,1]$ is exponentially tight and obeys the local large deviation principle in D, then the sequence X^n, $n \geq 1$, obeys the large deviation principle with the rate function $I = J$. (Here $J = J(\varphi)$ is the function from Definition 1.7.)

Proof. Assume that an exponentially tight sequence X^n, $n \geq 1$, satisfies the local large deviation principle. By Theorem 1.9 (see also Definition 1.6) one can choose a subsequence $X^{n''}$ which obeys the (full) large deviation principle with a rate function $\tilde{I}$. Then, by Theorem 1.7, the sequence $X^{n''}$ satisfies the local large deviation principle with rate function $\tilde{I}$. This means that

$$\begin{aligned}
J(\varphi) &= -\liminf_{\delta \downarrow 0} \liminf_{n \to \infty} n^{-1} \ln \mathsf{P}\,(\rho\,(X^n, \varphi) \leq \delta) \\
&\leq -\liminf_{\delta \downarrow 0} \liminf_{n'' \to \infty} \frac{1}{n''} \ln \mathsf{P}\,(\rho\,(X^{n''}, \varphi) \leq \delta) \\
&= -\limsup_{\delta \downarrow 0} \limsup_{n'' \to \infty} \frac{1}{n''} \ln \mathsf{P}\,(\rho\,(X^{n''}, \varphi) \leq \delta) = \tilde{I}(\varphi) \\
&\leq -\limsup_{\delta \downarrow 0} \limsup_{n \to \infty} n^{-1} \ln \mathsf{P}\,(\rho\,(X^n, \varphi) \leq \delta) = J(\varphi).
\end{aligned}$$

Hence, $\tilde{I} = J$ and J is a rate function.

It remains to prove that (X^n) satisfies conditions (I) and (II) of Definition 1.1. We will prove (I). (II) is proved similarly.

Let $G \subset D$ be an open set. By Theorem 1.9 one can choose a subsequence (n') of (n) such that

$$\liminf_n n^{-1} \ln \mathsf{P}(X^n \in G) = \lim_{n'} \frac{1}{n'} \ln \mathsf{P}(X^{n'} \in G) \tag{1.36}$$

and from (n') one can choose a subsequence (n'') such that $X^{n''}$ obeys the large deviation principle with rate function I''. Particularly, this means that

$$\liminf_{n''} \frac{1}{n''} \ln \mathsf{P}(X^{n''} \in G) \geq - \inf_{\varphi \in G} I''(\varphi). \tag{1.37}$$

However, we may choose (n'') as the sequence (n'') above. Then, $I'' = J$ and, by (1.36) and (1.37)

$$\liminf_n n^{-1} \ln \mathsf{P}(X^n \in G) \geq - \inf_{\varphi \in G} J(\varphi).$$

It completes the proof.

Theorem 1.10 suggests the following scheme for proving the large deviation principle:

Step 1. For $\varphi = (\varphi_t)$, $t \leq 1$, $\varphi_t \in D[0,1]$ calculate

$$\begin{aligned} J(\varphi) &= -\liminf_{\delta \downarrow 0} \liminf_{n \to \infty} n^{-1} \ln \mathsf{P}(\rho(X^n, \varphi) \leq \delta) \\ &= -\limsup_{\delta \downarrow 0} \limsup_{n \to \infty} n^{-1} \ln \mathsf{P}(\rho(X^n, \varphi) \leq \delta). \end{aligned} \tag{1.38}$$

Step 2. Check that the sequence (X^n), $n \geq 1$, is exponentially tight.

Step 3. Put $I(\varphi) = J(\varphi)$.

Theorem 1.10 gives the way to pass from local to full large deviation principle. However, it is not easy to deal with the Skorokhod metric. One would like to work with the uniform topology and have a constructive way for checking exponential tightness.

Definition 1.8. An exponentially tight sequence $X^n = (X_t^n)_{t \leq 1}$, $n \geq 1$, is said to be C-exponentially tight if for any subsequence $(X^{n'})$ which has the large deviation property with a rate function $I'(\varphi)$ the following property takes place: $I'(\varphi) = \infty$, for all $\varphi \in D[0,1] \backslash C[0,1]$.

Theorem 1.11. Let a sequence X^n, $n \geq 1$, be C-exponentially tight. Assume that for all $\varphi = (\varphi_t)_{t \leq 1}$, $\varphi_t \in C[0,1]$

$$\begin{aligned} J(\varphi) &= -\liminf_{\delta \downarrow 0} \liminf_{n \to \infty} n^{-1} \ln \mathsf{P}\left(\sup_{t \leq 1} |X_t^n - \varphi_t| \leq \delta\right) \\ &= -\limsup_{\delta \downarrow 0} \limsup_{n \to \infty} n^{-1} \ln \mathsf{P}\left(\sup_{t \leq 1} |X_t^n - \varphi_t| \leq \delta\right). \end{aligned} \tag{1.39}$$

Then X^n obeys the large deviation principle with the rate function

$$I(\varphi) = \begin{cases} J(\varphi), & \text{for } \varphi \in C[0,1], \\ \infty, & \text{for } \varphi \in D[0,1] \backslash C[0,1]. \end{cases}$$

Proof. By Theorem 1.10 and the definition of C-exponential tightness it suffices to prove that (1.39) implies (1.38). To this end, consider the Prokhorov complete metric (see, e.g., Billingsley (1969))

$$\rho(Y,Z) \equiv d_0(Y,Z) = \inf_{\lambda\in\Lambda}\left(\sup_{t\leq 1}\left|Y_t - Z_{\lambda(t)}\right| + \sup_{0\leq s\leq t\leq 1}\left|\ln\frac{\lambda(t)-\lambda(s)}{t-s}\right|\right), \tag{1.40}$$

where $Y = (Y_t)_{t\leq 1}$, $Z = (Z_t)_{t\leq 1}$, $Y, Z \in D[0,1]$ and Λ is the set of all strictly increasing continuous functions $\lambda = \lambda(t)$, $t \leq 1$, $\lambda(0) = 0$, $\lambda(1) = 1$. Evidently,

$$d_0(Y,Z) \leq \sup_{t\leq 1} |Y_t - Z_t|.$$

Hence, by (1.38) and (1.39), for all $\delta > 0$ and $\varphi \in C[0,1]$ we have:

$$\liminf_{n\to\infty} n^{-1}\ln \mathsf{P}\left(\rho(X^n,\varphi)\leq\delta\right) \geq \liminf_{n\to\infty} n^{-1}\ln \mathsf{P}\left(\sup_{t\leq 1}|X_t^n - \varphi_t| \leq \delta\right) \geq -J(\varphi).$$

To obtain upper bound we need the following result.

Lemma 1.4. If $\delta < 1/4$ then

$$(\rho(X,\varphi)\leq\delta) \subseteq \left(\sup_{t\leq 1}|X_t - \varphi_t| \leq 2\delta + W(\varphi, 4\delta)\right), \tag{1.41}$$

where $(0 \leq s,\, u \leq 1)$

$$W(\varphi, h) = \sup_{|s-u|\leq h} |\varphi_s - \varphi_u|. \tag{1.42}$$

Proof of Lemma 1.4. Note (see Billingsley (1969) Chapter 3, 14) that if

$$\sup_{0\leq s\leq t\leq 1}\left|\ln\frac{\lambda(t)-\lambda(s)}{t-s}\right| \leq 1/4,$$

then

$$\frac{1}{2}\sup_{t\leq 1}|\lambda(t) - t| \leq \sup_{0\leq s\leq t\leq 1}\left|\ln\frac{\lambda(t)-\lambda(s)}{t-s}\right|.$$

Therefore, if $\rho(X,\varphi) \leq 1/4$ then

$$\rho(X,\varphi) \geq \inf_{\lambda\in\Lambda}\left(\sup_{t\leq 1}|X_t - \varphi_{\lambda_\delta(t)}| + \frac{1}{2}\sup_{t\leq 1}|\lambda_\delta(t) - t|\right)$$

and so for any $\delta \leq 1/4$ there exists $\lambda_\delta \in \Lambda$ such that

$$\rho(X,\varphi) \geq \sup_{t\leq 1}|X_t - \varphi_{\lambda_\delta(t)}| + \frac{1}{2}\sup_{t\leq 1}|\lambda_\delta(t) - t| - \delta. \tag{1.43}$$

Hence, if $\rho(X,\varphi) \leq \delta \leq 1/4$ then

$$\sup_{t\leq 1}|\lambda_\delta(t) - t| \leq 4\delta$$

and so, by (1.42) for $0 \leq s,\, u \leq 1$

$$\sup_{t\leq 1}|\varphi_t - \varphi_{\lambda_\delta(t)}| \leq \sup_{|s-u|\leq 4\delta}|\varphi_s - \varphi_u| = W(\varphi, 4\delta). \tag{1.44}$$

We next note that by the triangle inequality

$$\sup_{t\le 1} |X_t - \varphi_{\lambda_\delta(t)}| \ge \sup_{t\le 1} |X_t - \varphi_t| - \sup_{t\le 1} |\varphi_t - \varphi_{\lambda_\delta(t)}|.$$

Then, from (1.43) we get

$$\rho(X,\varphi) \ge \sup_{t\le 1} |X_t - \varphi_t| - \sup_{t\le 1} |\varphi_t - \varphi_{\lambda_\delta(t)}| - \delta$$

and so, by (1.44)

$$\rho(X,\varphi) \ge \sup_{t\le 1} |X_t - \varphi_t| - W(\varphi, 4\delta) - \delta,$$

which gives (1.41). Lemma 1.4 is proved.

Since $\varphi_t \in C[0,1]$, it follows from (1.42) that $W(\varphi, 4\delta) \to 0$ as $\delta \to 0$ and then by (1.41) and (1.39)

$$\limsup_{\delta\downarrow 0} \limsup_{n\to\infty} n^{-1} \ln \mathsf{P}\,(\rho\,(X^n,\varphi) \le \delta)$$
$$\le \limsup_{\delta\downarrow 0} \limsup_{n\to\infty} n^{-1} \ln \mathsf{P}\left(\sup_{t\le 1} |X_t^n - \varphi_t| \le \delta\right) \le -J(\varphi).$$

It completes the proof of theorem.

To give the practical way for checking of exponential tightness we have to restrict further the processes under consideration to the class of adapted processes.

Recall that stochastic process $X = (X_t)_{0\le t\le 1}$, defined on a stochastic basis $(\Omega, F, \mathbf{F}, \mathsf{P})$ with a filtration $\mathbf{F} = (F_t)_{t\le 1}$ is said to be $\mathbf{F}$-adapted if, for each $t \le 1$, the variables X_t are F_t-measurable.

Recall also that a random variable $\tau = \tau(\omega)$ is said to be a stopping time relative to a filtration $\mathbf{F}$ if, for every $t \le 1$, the set $(\omega\colon \tau(\omega) \le t)$ is F_t-measurable.

Theorem 1.12. Let $X^n = (X_t^n)_{0\le t\le 1}$, $n \ge 1$, be a sequence of processes with path in $D[0,1]$. Assume that each X^n is defined on a stochastic basis $(\Omega, F, \mathbf{F}^n, \mathsf{P})$ with a filtration $\mathbf{F}^n = (F_t^n)_{t\le 1}$. Assume that

$$\lim_{c\to\infty} \limsup_{n} n^{-1} \ln \mathsf{P}\left(\sup_{t\le 1} |X_t^n| > c\right) = -\infty \tag{i}$$

and for all $\eta > 0$

$$\limsup_{\delta\to 0} \limsup_{n} \sup_{\tau\in T(\mathbf{F}^n)} n^{-1} \ln \mathsf{P}\left(\sup_{t\le\delta} |X_{\tau+t}^n - X_\tau^n| \ge \eta\right) = -\infty, \tag{ii}$$

where $T(\mathbf{F}^n)$ is the set of stopping time relative to a filtration $\mathbf{F}^n$.

Then the sequence X^n, $n \ge 1$, is exponentially tight and C-exponentially tight.

Proof. It is proved by Pukhalskii (1991) that (i) and (ii) imply exponential tightness of the sequence X^n, $n \ge 1$, and (ii) also implies C-exponential tightness. We adopt here only the second statement.

Let $\varphi \in D[0,1]\backslash C$. Then there exists $s_0 \in (0,1]$ with $\Delta\varphi_{s_0} \neq 0$. (For $X \in D[0,1]$ we denote $\Delta X = X_t - X_{t-}$, $X_{t-} = \lim_{s\uparrow t} X_s$, $t \leq 1$.)

Lemma 1.5. If $\delta \leq 1/4$ then

$$(\rho(Y,\varphi) \equiv d_0(Y,\varphi) \leq \delta) \subseteq \left(\sup_{|s_0 - t| \leq 4\delta} |Y_{s_0} - Y_t| \geq |\Delta\varphi_{s_0}| - 2\delta \right). \tag{1.45}$$

Proof of Lemma 1.5. We follow the outline of the proof of Lemma 1.4 (see the proof of Theorem 1.11 in this section). If $d_0(Y,\varphi) \leq \delta \leq 1/4$ one can choose $\lambda_\delta = \lambda_\delta(t)$, $\lambda_\delta \in \Lambda$, such that

$$d_0(Y,\varphi) \geq \sup_{t \leq 1} |\varphi_t - Y_{\lambda_\delta(t)}| + \frac{1}{2} \sup_{t \leq 1} |\lambda_\delta(t) - t| - \delta,$$

and

$$\sup_{t \leq 1} |\lambda_\delta(t) - t| \leq 4\delta.$$

Hence,

$$|\Delta Y_{\lambda_\delta(s_0)}| = |Y_{\lambda_\delta(s_0)} - Y_{\lambda-}| \leq 2 \sup_{|t - s_0| \leq 4\delta} |Y_{s_0} - Y_t| . \tag{1.46}$$

where $Y_{\lambda-} = \lim_{t \uparrow \lambda_\delta(s_0)} Y_t$. We next note that by the triangle inequality

$$\begin{aligned} |\Delta\varphi_{s_0}| &\leq |\Delta Y_{\lambda_\delta(s_0)} - \Delta\varphi_{s_0}| + |\Delta Y_{\lambda_\delta(s_0)}| \\ &\leq 2 \sup_{t \leq 1} |\varphi_t - Y_{\lambda_\delta(t)}| + |\Delta Y_{\lambda_\delta(s_0)}|. \end{aligned} \tag{1.47}$$

By (1.46) and (1.47), if $d_0(Y,\varphi) \leq \delta \leq 1/4$ then

$$\begin{aligned} \sup_{|t - s_0| \leq 4\delta} |Y_{s_0} - Y_t| &\geq \frac{1}{2} |\Delta\varphi_{s_0}| - \sup_{t \leq 1} |\varphi_t - Y_{\lambda_\delta(t)}| \\ &\geq \frac{1}{2} |\Delta\varphi_{s_0}| - (d_0(Y,\varphi) + \delta), \end{aligned}$$

which gives (1.45). This completes the proof of Lemma 1.5.

Evidently, since $s_0 < 1$

$$\mathsf{P}\left(\sup_{|t - s_0| \leq 4\delta} |X^n_{s_0} - X^n_t| \geq \eta \right) \leq 2 \sup_{\tau \in T(\mathbf{F}^n)} \mathsf{P}\left(\sup_{s \leq 4\delta} |X^n_{\tau+s} - X^n_\tau| \geq \eta \right).$$

Therefore, using (1.45) and (ii) we conclude that, for $\varphi \in D[0,1]\backslash C[0,1]$

$$\begin{aligned} &\limsup_{\delta} \limsup_{n} n^{-1} \ln \mathsf{P}(\rho(X^n,\varphi) \leq \delta) \\ &\leq \limsup_{\delta \to 0} \limsup_{n} n^{-1} \ln \left(2 \sup_{\tau \in T(\mathbf{F}^n)} \mathsf{P}\left(\sup_{s \leq 4\delta} |X^n_{\tau+s} - X^n_\tau| \right.\right. \\ &\geq \left.\left. \frac{|\Delta\varphi_{s_0}|}{2} - 2\delta \right) \right) = -\infty. \end{aligned}$$

Now assume that a subsequence $X^{n'}$ satisfies the large deviation principle with a rate function I'. Then, since it obeys the local large deviation principle, from the preceding we have that $I' = \infty$, for $\varphi \in D[0,1]\backslash C[0,1]$.

This completes the proof of Theorem 1.12.

Thus, we arrive at the following scheme for proving the large deviation principle once X^n is the sequence of adapted processes with path in $D[0,1]$:

Step 1. For $\varphi = (\varphi_t)_{t\geq 0}$, $\varphi_t \in C[0,1]$ calculate

$$\begin{aligned} J(\varphi) &= -\liminf_{\delta\downarrow 0} \liminf_{n\to\infty} n^{-1} \ln \mathsf{P}\left(\sup_{t\leq 1} |X_t^n - \varphi_t| \leq \delta\right) \\ &= -\limsup_{\delta\downarrow 0} \limsup_{n\to\infty} n^{-1} \ln \mathsf{P}\left(\sup_{t\leq 1} |X_t^n - \varphi_t| \leq \delta\right). \end{aligned} \tag{1.48}$$

Step 2. Check that the sequence (X^n), $n \geq 1$, is C-exponentially tight, i.e.

$$\lim_{c\to\infty} \limsup_{n} n^{-1} \ln \mathsf{P}\left(\sup_{t\leq 1} |X_t^n| > c\right) = -\infty \tag{1.49}$$

and for all $\eta > 0$

$$\limsup_{\delta\to 0} \limsup_{n} \sup_{\tau\in T(\mathbf{F}^n)} n^{-1} \ln \mathsf{P}\left(\sup_{t\leq\delta} |X_{\tau+t}^n - X_\tau^n| \geq \eta\right) = -\infty, \tag{1.50}$$

Step 3. Put

$$I(\varphi) = \begin{cases} J(\varphi), & \text{for } \varphi \in C[0,1], \\ \infty, & \text{for } \varphi \in D[0,1]\backslash C[0,1]. \end{cases}$$

1.5. LARGE DEVIATIONS FOR SUMS OF INDEPENDENT AND IDENTICALLY DISTRIBUTED VARIABLES IN FUNCTION SPACE

In this section we consider the simplest example of the large deviation result in a function space. In our treatment we follow the scheme of Section 1.4.

Setting

Let $(\xi_k)_{k\geq 1}$ be a sequence of i.i.d. random variables with a common distribution μ, $E\xi_1 = 0$, that are defined on a probability space (Ω, F, P) and take values in R. Define $(F_k)_{k\geq 1}$ a family of σ-fields, with F_k the σ-field, generated by $\xi_1, \ldots, \xi_k$.

Assume that *the moment generating function* (Cramér's condition)

$$\mathsf{E} \exp(\lambda\xi) \tag{1.51}$$

is finite for all $\lambda \in R$ and *the logarithmic moment generating function*

$$G_\mu(\lambda) = \ln \mathsf{E} \exp(\lambda\xi)$$

satisfies the regularity condition

$$\lim_{|\lambda|\to\infty} \frac{G_\mu(\lambda)}{|\lambda|} = \infty. \tag{R}$$

Introduce the family of processes $W^n = (W_t^n)_{0\le t\le 1}$ with

$$W_t^n = n^{-1}\sum_{k=1}^{[nt]} \xi_k \tag{1.52}$$

(recall that $[s]$ is the integer part of $s \in R$). Clearly, W^n converges as $n \to \infty$ to the trajectory $x(t)$, which never leaves 0.

Let φ be a function on the closed interval $[0, 1]$ with values in R which vanish at the origin. The problem in which we interest is that of determining an asymptotic behavior of the probability

$$\mathsf{P}\left(\sup_{t\le 1} |W_t^n - \varphi(t)| \le \delta\right), \qquad \delta > 0,$$

i.e., what is the probability that the trajectory W_t^n be close in the time interval $[0, 1]$ to a preassigned function φ_t?

In this section we will prove the following result.

Theorem 1.13. Let Cramér's condition and the regularity condition (R) hold. Then the sequence W^n, $n \ge 1$, obeys the large deviation principle in $D[0, 1]$ with the rate function

$$I(\varphi) = \begin{cases} \int_0^1 L(\dot\varphi(s))\, ds, & \varphi \text{ is absolutely continuous}, \quad \varphi_0 = 0, \\ \infty, & \text{otherwise}, \end{cases} \tag{1.53}$$

where

$$L(x) = \sup_{\lambda\in R} [\lambda x - G_\mu(\lambda)].$$

Remark 1.3. Assume that (ξ_k) are Gaussian with $\mathsf{E}\xi_1^2 = 1$ then

$$G_\mu(\lambda) = \ln \mathsf{E} \exp(\lambda\xi_1) = \lambda^2/2$$

and

$$I(\varphi) = \begin{cases} \int_0^1 (\dot\varphi(s))^2\, ds, & \varphi \text{ is absolutely continuous}, \quad \varphi_0 = 0, \\ \infty, & \text{otherwise}. \end{cases}$$

Transformation of the measure

As in Section 1.1, we will exploit Cramér's idea of changing the given measure. Let $\lambda = (\lambda_k)$, $k \ge 0$, be a numerical sequence with $\lambda_k \in R$. Define

$$z^n(\lambda) = \exp\left(\sum_{k=1}^n \lambda_k\xi_k - \sum_{k=1}^n G_\mu(\lambda_k)\right). \tag{1.54}$$

It easy to see that

$$\mathsf{E}z^n(\lambda) = 1. \tag{1.55}$$

This equality enables us to define a new probability measure $Q^{n,\lambda}$ by

$$dQ^{n,\lambda} = z^n(\lambda)\, d\mathsf{P}. \tag{1.56}$$

Since we are going to establish a large deviations property in infinite dimensional case it is natural to use the following definitions and notations. Let $\lambda(s)$, $0 \le s \le 1$, be a continuous function with

$$\lambda\left(\frac{k}{n}\right) = \lambda_k.$$

Introduce an increasing process

$$A_t^n = \frac{[nt]}{n}, \qquad 0 \le t \le 1,$$

and define $z_t^n(\lambda)$ as

$$z_t^n(\lambda) = \exp\left(n\left[\int_0^t \lambda(s)\, \mathrm{d}W_s^n - \int_0^t G_\mu(\lambda(s))\, \mathrm{d}A_s^n\right]\right), \tag{1.57}$$

where ($\Delta = 1/n$)

$$\begin{aligned} \int_0^t G_\mu(\lambda(s))\, \mathrm{d}A_s^n &= n^{-1}\sum_{k=1}^{[nt]} G_\mu(\lambda(k\Delta)), \\ \int_0^t \lambda(s)\, \mathrm{d}W_s^n &= n^{-1}\sum_{k=1}^{[nt]} \lambda(k\Delta)\xi_k. \end{aligned} \tag{1.58}$$

Note that $z_1^n(\lambda) \equiv z^n(\lambda)$, where $z^n(\lambda)$ is defined by (1.54).

One can see that, by definition, $Q^{n,\lambda}$ is a product-measure and consequently w.r.t. the measure $Q^{n,\lambda}$, (ξ_k) is the sequence of independent but not identically distributed random variables with $\mathsf{E}^{n,\lambda}\xi_k \neq 0$ ($\mathsf{E}^{n,\lambda}$ denotes the expectation w.r.t. $Q^{n,\lambda}$).

Later on the value of the $\mathsf{E}^{n,\lambda}\xi_k$ plays an important role and so we now calculate it. By independence of (ξ_k) w.r.t. $Q^{n,\lambda}$, taking into account (1.54) and (1.56), we get

$$\begin{aligned} \mathsf{E}^{n,\lambda}\xi_k &= \mathsf{E}\xi_k \frac{\exp\left(n\int_0^1 \lambda(s)\, \mathrm{d}W_s^n\right)}{\exp\left(n\int_0^1 G_\mu(\lambda(s))\, \mathrm{d}A_s^n\right)} \\ &= \frac{\mathsf{E}\xi_k \exp(\lambda(k\Delta)\xi_k)\prod_{i\neq k}\exp(\lambda(i\Delta)\xi_i)}{\mathsf{E}\exp(\lambda(k\Delta)\xi_k)\prod_{i\neq k}\exp(\lambda(i\Delta)\xi)} \\ &= \frac{\mathsf{E}(\xi_k \exp(\lambda(k\Delta)\xi_k))}{\mathsf{E}\exp(\lambda(k\Delta)\xi_k)}. \end{aligned} \tag{1.59}$$

Note, that at the same time

$$\left.\frac{\mathrm{d}G_\mu(\lambda)}{\mathrm{d}\lambda}\right|_{\lambda=\lambda(k\Delta)} = \frac{\mathsf{E}(\xi_k \exp(\lambda(k\Delta)\xi_k))}{\mathsf{E}\exp(\lambda(k\Delta)\xi_k)}. \tag{1.60}$$

Hence,

$$\mathsf{E}^{n,\lambda}\xi_k = \left.\frac{dG_\mu(\lambda)}{d\lambda}\right|_{\lambda=\lambda(k\Delta)} \quad (= G'_\mu(\lambda(k\Delta))). \tag{1.61}$$

We now introduce the following decomposition of W_t^n with respect to the measure $Q^{n,\lambda}$:

$$W_t^n = B_t^n(\lambda) + M_t^n(\lambda) \tag{1.62}$$

with

$$\begin{aligned} M_t^n(\lambda) &= n^{-1}\sum_{k=1}^{[nt]}(\xi_k - \mathsf{E}^{n,\lambda}\xi_k) &= n^{-1}\sum_{k=1}^{[nt]}[\xi_k - G'_\mu(\lambda(k\Delta))], \\ B_t^n(\lambda) &= n^{-1}\sum_{k=1}^{[nt]}\mathsf{E}^{n,\lambda}\xi_k &= \int_0^t G'_\mu(\lambda(s))\, dA_s^n. \end{aligned} \tag{1.63}$$

Clearly, $M_t^n(\lambda)$ is the sum of independent w.r.t $Q^{n,\lambda}$ zero mean random variables

$$\xi_k - G'_\mu(\lambda(k\Delta)), \qquad k \geq 1.$$

In the transformation (1.57) the function $\lambda(s)$ is not yet fixed. Consider a continuously differentiable function $\varphi(t)$ with $\varphi(0) = 0$, $0 \leq t \leq 1$, and denote $d\varphi(t)/dt = \dot\varphi(t)$. We choose $\lambda^0(s) = \lambda(\dot\varphi(s))$ as the solution of the equation

$$\frac{d}{d\lambda}(\lambda\dot\varphi(s) - G_\mu(\lambda)) = 0. \tag{1.64}$$

Under the regularity condition (R) the supremum

$$L(y) = \sup_y\,(\lambda y - G_\mu(\lambda)) \tag{1.65}$$

is attained at some λ for the given $y = \dot\varphi(s)$ and the solution $\lambda(\dot\varphi(s))$ is a continuous bounded function such that

$$G'_\mu(\lambda^0(s)) = \dot\varphi(s). \tag{1.66}$$

Define

$$\int_0^t \dot\varphi(s)\, dA_s^n = n^{-1}\sum_{k=1}^{[nt]}\dot\varphi(k\Delta). \tag{1.67}$$

Taking into account (1.63), (1.66) and (1.67), we write (1.62) in the following form:

$$\begin{aligned} W_t^n &= \int_0^t G'_\mu(\lambda(\dot\varphi(s)))\, dA_s^n + M_t^n(\lambda(\dot\varphi(s))) \\ &= \int_0^t \dot\varphi(s)\, dA_s^n + M_t^n(\lambda(\dot\varphi(s))). \end{aligned}$$

Hence,

$$M_t^n(\lambda(\dot\varphi(s))) = W_t^n - \int_0^t \dot\varphi(s)\, ds + i_1^n,$$

where

$$i_1^n = \int_0^t \dot{\varphi}(s)\, \mathrm{d}(s - A_s^n).$$

By the definition $A_s^n = ([ns]/n) \to s$, as $n \to \infty$, and we may expect that as $n \to \infty$, $i_1^n \to 0$. Hence, one can see that the idea of the transformation is to make $W_t^n - \int_0^t \dot{\varphi}(s)\, \mathrm{d}s = W_t^n - \varphi(t)$ be close to the $M_t^n(\lambda(\dot{\varphi}(s)))$.

With these preliminaries we are ready to formulate the first result.

Lower bound

Theorem 1.14. Let $\varphi(s)$, $0 \le s \le 1$, be a continuously differentiable function with $\varphi(0) = 0$. If Cramér's condition and the regularity condition (R) hold then

$$\liminf_{\delta \to 0} \liminf_{n \to \infty} \ln \mathrm{P}\left(\sup_{t \le 1} |W_t^n - \varphi(t)| \le \delta\right) \ge -I(\varphi) \tag{1.68}$$

with

$$I(\varphi) = \int_0^1 L(\dot{\varphi}(s))\, \mathrm{d}s.$$

Proof. Fix $\lambda^0(s)$ satisfying (1.64). Since under Cramér's hypothesis $z^n(\lambda^0) > 0$, we have

$$\mathrm{dP} = [z^n(\lambda^0)]^{-1}\, \mathrm{d}Q^{n,\lambda^0}.$$

Denote (see (1.63))

$$\int_0^1 \lambda^0(s)\, \mathrm{d}M_s^n(\lambda^0) = n^{-1} \sum_{k=1}^n \lambda^0(k\Delta)(\xi_k - G'_\mu(\lambda^0(k\Delta))). \tag{1.69}$$

Combining (1.57), (1.62), (1.63) and (1.69) we get

$$\begin{aligned}
(z^n(\lambda^0))^{-1} &= \exp\left(-n\left(\int_0^1 \lambda^0(s)\, \mathrm{d}W_s^n - \int_0^1 G_\mu(\lambda^0(s))\, \mathrm{d}A_s^n\right)\right)\\
&= \exp\left(-n\left(\int_0^1 \lambda^0(s)\, \mathrm{d}M_s^n(\lambda^0) + \int_0^1 \lambda^0(s)\, \mathrm{d}B_s^n(\lambda^0)\right.\right.\\
&\qquad \left.\left. - \int_0^1 G_\mu(\lambda^0(s))\, \mathrm{d}A_s^n\right)\right)\\
&= \exp\left(-n\left(\int_0^1 \lambda^0(s)\, \mathrm{d}M_s^n(\lambda^0) + \int_0^1 \lambda^0(s) G'_\mu(\lambda^0(s))\, \mathrm{d}A_s^n\right.\right.\\
&\qquad \left.\left. - \int_0^1 G_\mu(\lambda^0(s))\, \mathrm{d}A_s^n\right)\right).
\end{aligned} \tag{1.70}$$

Then, taking into account (1.66), we write

$$\mathsf{P}\left(\sup_{t\le 1}|W_t^n-\varphi(t)|\le\delta\right)=\int \mathrm{I}\left(\sup_{t\le 1}|W_t^n-\varphi(t)|\le\delta\right)\times\exp\left(-n\left(\int_0^1\lambda^0(s)\,\mathrm{d}M_s^n(\lambda^0)+\int_0^1(\lambda^0(s)\dot\varphi(s)-G_\mu(\lambda^0(s)))\,\mathrm{d}A_s^n\right)\right)\mathrm{d}Q^{n,\lambda^0}. \tag{1.71}$$

Denote

$$J_1^n=\int_0^1(\lambda^0(s)\dot\varphi(s)-G_\mu(\lambda^0(s)))\,\mathrm{d}A_s^n-\int_0^1(\lambda^0(s)\dot\varphi(s)-G_\mu(\lambda^0(s)))\,\mathrm{d}s, \tag{1.72}$$

and

$$C_{\delta,q}=\left(\sup_{t\le 1}|W_t^n-\varphi(t)|\le\delta\right)\bigcap\left(\left|\int_0^1\lambda^0(s)\,\mathrm{d}M_s^{n,\lambda^0}\right|\le q\right) \tag{1.73}$$

with some $q>0$. Thus, by (1.71)-(1.73), taking into account (1.65), we arrive at the following inequality:

$$\begin{aligned}\mathsf{P}&\left(\sup_{t\le 1}|W_t^n-\varphi(t)|\le\delta\right)\\ &\ge\int_{C_{\delta,q}}\exp\left(-n\left[\int_0^1\lambda^0(s)\,\mathrm{d}M_s^n(\lambda^0)+\int_0^1L(\dot\varphi(s))\,\mathrm{d}s+J_1^n\right]\right)\mathrm{d}Q^{n,\lambda^0}\\ &\ge\int_{C_{\delta,q}}\exp\left(-n\left[\int_0^1L(\dot\varphi(s))\,\mathrm{d}s+q+J_1^n\right]\right)\mathrm{d}Q^{n,\lambda^0},\end{aligned} \tag{1.74}$$

and so

$$n^{-1}\ln\mathsf{P}\left(\sup_{t\le 1}|W_t^n-\varphi(t)|\le\delta\right)\ge -I(\varphi)-q-J_1^n+n^{-1}\ln Q^{n,\lambda^0}(C_{\delta,q}).$$

Since q is arbitrary, the statement of Theorem 1.14 will be proved once we show that

$$\lim_{n\to\infty}J_1^n=0 \tag{1.75}$$

and

$$\lim_{n\to\infty}Q^{n,\lambda^0}(C_{\delta,q})=1$$

for every $q>0$ and $\delta>0$. Note that

$$Q^{n,\lambda^0}(C_{\delta,q})\ge 1-Q^{n,\lambda^0}\left(\sup_{t\le 1}|W_t^n-\varphi(t)|>\delta\right)-Q^{n,\lambda^0}\left(\left|\int_0^1\lambda^0(s)\,\mathrm{d}M_s^n(\lambda^0)\right|>q\right).$$

Hence, we must prove that

$$\lim_{n\to\infty}Q^{n,\lambda^0}\left(\sup_{t\le 1}|W_t^n-\varphi(t)|>\delta\right)=0 \tag{1.76}$$

for all $\delta > 0$ and

$$\lim_{n\to\infty} Q^{n,\lambda^0}\left(\left|\int_0^1 \lambda^0(s)\, dM_s^n(\lambda^0)\right| > q\right) = 0 \tag{1.77}$$

for every $q > 0$. Here and in Chapter 2 we will use the following simple fact.

Lemma 1.6. Let $u(s)$ be a continuous function. Then

$$\lim_{n\to\infty} \int_0^1 u(s)\, dA_s^n = \int_0^1 u(s)\, ds, \tag{1.78}$$

$$\lim_{n\to\infty} \sup_{0\le t\le 1} \left|\int_0^t u(s)\, dA_s^n - \int_0^t u(s)\, ds\right| = 0. \tag{1.79}$$

Proof. Note that (1.78) means simply a weak convergence of the measure dA_s^n as $n \to \infty$ to Lebesque measure on $[0, 1]$. So (1.78) follows from the definition of A_t^n. To prove (1.79) we use the following representation:

$$\int_0^t u(s)\, dA_s^n = \int_0^t u^+(s)\, dA_s^n - \int_0^t u^-(s)\, dA_s^n.$$

By repeating the argument of weak convergence, we obtain

$$\lim_{n\to\infty} \int_0^t u^+(s)\, dA_s^n = \int_0^t u^+(s)\, ds$$

and

$$\lim_{n\to\infty} \int_0^t u^-(s)\, dA_s^n = \int_0^t u^-(s)\, ds$$

for any $t \in [0, 1]$. Thus, by Polya's theorem we get the uniform convergence. Lemma is proved.

Armed with these results, we immediately obtain (1.75), by observing that

$$u(s) = \lambda^0(s)\dot{\varphi}(s) - G_\mu(\lambda^0(s))$$

is a continuous function.

Turning to the proof of (1.76), we note that, in virtue of (1.62),

$$\begin{aligned}\sup_{0\le t\le 1} |W_t^n - \varphi(t)| &= \sup_{0\le t\le 1} |M_t^n(\lambda^0) + B_t^n(\lambda^0) - \varphi(t)| \\ &\le \sup_{0\le t\le 1} |M_t^n(\lambda^0)| + \sup_{0\le t\le 1} |B_t^n(\lambda^0) - \varphi(t)|.\end{aligned}$$

We next apply (1.79) to obtain, taking into account (1.63) and (1.66), that

$$\lim_{n\to\infty}\sup_{0\le t\le 1}|B^n_t(\lambda^0)-\varphi(t)| = \lim_{n\to\infty}\sup_{0\le t\le 1}\left|\int_0^t G'_\mu(\lambda^0(s))\,\mathrm{d}A^n_s-\int_0^t\dot\varphi(s)\,\mathrm{d}s\right|$$
$$= \lim_{n\to\infty}\sup_{0\le t\le 1}\left|\int_0^t \dot\varphi(s)\,\mathrm{d}A^n_s-\int_0^t\dot\varphi(s)\,\mathrm{d}s\right| = 0.$$

Hence, it remains to prove (1.77) and that

$$\lim_{n\to\infty} Q^{n,\lambda^0}\left(\sup_{0\le t\le 1}|M^n_t(\lambda^0)|>\delta\right)=0 \tag{1.80}$$

for every $\delta>0$.

In both these cases we are dealing with the sums of independent variables. So in proving (1.77) and (1.80) we can use Chebyshov's and Kolmogorov's inequalities. As a preliminary to these estimates, we estimate the variance ξ_k w.r.t. Q^{n,λ^0}:

$$\mathrm{E}^{n,\lambda^0}\xi_k^2 = \mathrm{E}\xi_k^2 z^n(\lambda^0) = \mathrm{E}\xi_k^2\frac{\exp(\lambda^0(k\Delta)\xi_k)}{\exp(G_\mu(\lambda^0(k\Delta)))}$$
$$\le \mathrm{E}\xi_1^2\frac{\exp\left(\sup_{0\le s\le 1}|\lambda^0(s)|\xi_1\right)}{\exp\left(-\sup_{0\le s\le 1}|G_\mu(\lambda^0(s))|\right)}.$$

Note, that under Cramér's condition the right-hand side of the last inequality is finite.

Turning to prove of (1.77) we apply Chebyshov's inequality to show that

$$Q^{n,\lambda^0}\left(\left|\int_0^1\lambda^0(s)\,\mathrm{d}M^n_s(\lambda^0)\right|>q\right)$$
$$\le q^{-2}\mathrm{E}^{n,\lambda^0}\left(n^{-1}\sum_{k=1}^n\lambda^0(k\Delta)[\xi_k-G'_\mu(\lambda^0(k\Delta))]\right)^2$$
$$\le q^{-2}n^{-2}\sup_{0\le s\le 1}|\lambda^0(s)|^2\sum_{k=1}^n\mathrm{E}^{n,\lambda^0}\xi_k^2.$$

From the preceding, it is then an easy step to the conclusion that right-hand side of the inequality tends to zero as $n\to\infty$.

To prove (1.80) we note that by Kolmogorov's inequality

$$Q^{n,\lambda^0}\left(\sup_{0\le t\le 1}|M^n_t(\lambda^0)|>\delta\right) = Q^{n,\lambda^0}\left(\sup_{0\le t\le 1}\left|n^{-1}\sum_{k=1}^{[nt]}\xi_k-G'_\mu(\lambda^0(k\Delta))\right|\ge\delta\right)$$
$$\le 4\delta^{-2}\mathrm{E}^{n,\lambda^0}\left(n^{-1}\sum_{k=1}^n[\xi_k-G'_\mu(\lambda^0(k\Delta))]\right)^2$$
$$\le 4\delta^{-2}n^{-2}\sum_{k=1}^n\mathrm{E}^{n,\lambda^0}\xi_k^2.$$

By the same argument, we conclude that the right-hand side of the inequality tends to zero and (1.80) holds. Theorem 1.14 is proved.

Upper bound

Theorem 1.15. Under Cramér's condition, for all $\varphi \in D[0,1]$, $\varphi_0 = 0$,

$$\limsup_{\delta\to 0} \limsup_{n} n^{-1} \ln \mathrm{P}\left(\sup_{t\le 1} |W_t^n - \varphi(t)| \le \delta\right) \le -\tilde{I}(\varphi),$$

with

$$\tilde{I}(\varphi) = \sup \left[\int_0^1 \lambda(s)\, \mathrm{d}\varphi(s) - \int_0^1 G_\mu(\lambda(s))\, \mathrm{d}s\right], \tag{1.81}$$

where sup is taken over all step-functions λ.

Proof. Consider a step-function $(\lambda(t))_{t\le 1}$ of the form

$$\lambda(t) = \sum_i \lambda_{t_i} \mathrm{I}_{]t_{i-1},t_i]},$$

and define

$$\int_0^1 \lambda(t)\, \mathrm{d}\varphi(t) = \sum_i \lambda_{t_i}[\varphi_{1\wedge t_i} - \varphi_{1\wedge t_{i-1}}].$$

Recall that by definition (1.58), for any function $\lambda(s)$ and $\Delta = 1/n$

$$\int_0^1 \lambda(s)\, \mathrm{d}W_s^n = n^{-1} \sum_{k=1}^{[nt]} \lambda(k\Delta)\xi_k.$$

Consider $z^n(\lambda)$, defined by (1.57) with a step-function $\lambda(t)$ and recall that $\mathrm{E}z^n(\lambda) = 1$. Then, for $\varphi \in D[0,1]$,

$$\begin{aligned}
1 &\ge \mathrm{E\,I}\left(\sup_{t\le 1} |W_t^n - \varphi(t)| \le \delta\right) \exp\left(n\left[\int_0^1 \lambda(s)\, \mathrm{d}W_s^n - \int_0^1 G_\mu(\lambda(s))\, \mathrm{d}A_s^n\right]\right) \\
&\ge \mathrm{E\,I}\left(\sup_{t\le 1} |W_t^n - \varphi(t)| \le \delta\right) \exp\left(n\left[\int_0^1 \lambda(s)\, \mathrm{d}\varphi(s) - \int_0^1 G_\mu(\lambda(s))\, \mathrm{d}s\right.\right. \\
&\quad \left.\left. + \int_0^1 \lambda(s)\, \mathrm{d}(W_s^n - \varphi(s)) - \int_0^1 G_\mu(\lambda(s))\, \mathrm{d}(A_s^n - s)\right]\right) \\
&\ge \exp\left(n\left[\int_0^1 \lambda(s)\, \mathrm{d}\varphi(s) - \int_0^1 G_\mu(\lambda(s))\, \mathrm{d}s\right]\right) \\
&\quad \times \mathrm{E\,I}\,(\sup_{t\le 1} |W_t^n - \varphi(t)| \le \delta) \exp(n[-|i_1| - |i_2|]),
\end{aligned} \tag{1.82}$$

where

$$i_1 = \int_0^1 \lambda(s)\, \mathrm{d}(W_s^n - \varphi(s)),$$

$$i_2 = \int_0^1 G_\mu(\lambda(s))\, \mathrm{d}(A_s^n - s).$$

By Lemma 1.6, $|i_1| \to 0$ as $n \to \infty$. Then, for any $\varepsilon > 0$ one can choose $\delta_\varepsilon > 0$ such that on the set $(\sup_{t\le 1} |W_t^n - \varphi(t)| \le \delta)$ with $\delta < \delta_\varepsilon$

$$|i_1| + |i_2| \le \varepsilon,$$

and so from (1.82) we obtain

$$n^{-1} \ln \mathsf{P}\left(\sup_{t\le 1} |W_t^n - \varphi(t)| \le \delta\right) \le -\tilde{I}(\varphi) + \varepsilon.$$

Since ε is arbitrary, we get the statement of Theorem 1.15.

Exponential tightness

According to our scheme, we have to check the conditions (1.49) and (1.50). We first check that

$$\lim_{c\to\infty} \limsup_n n^{-1} \ln \mathsf{P}\left(\sup_{t\le 1} |W_t^n| > c\right) = -\infty.$$

To this end, define the stopping time

$$\theta_c = \inf\ (t\colon\ |W_t^n| \ge c) \wedge 1,$$

and note that

$$\left(\omega\colon\ \sup_{t\le 1} |W_t^n| \ge c\right) = (\omega\colon\ |W_{\theta_c}^n| \ge c) = (W_{\theta_c}^n \ge c) \cup (W_{\theta_c}^n \le -c).$$

Hence,

$$\mathsf{P}\left(\sup_{t\le 1} |W_t^n| \ge c\right) \le 2 \max\left[\mathsf{P}(W_{\theta_c}^n \ge c), \mathsf{P}(W_{\theta_c}^n \le -c)\right].$$

Therefore, it is sufficient to prove that

$$\begin{aligned} &\lim_{c\to\infty} \limsup_n n^{-1} \ln \mathsf{P}(W_{\theta_c}^n \ge c) \\ &\quad = \lim_{c\to\infty} \limsup_n n^{-1} \ln \mathsf{P}(W_{\theta_c}^n \le -c) = -\infty. \end{aligned} \tag{1.83}$$

To this end, note that the process $z_t^n(\lambda)$ (see (1.57) is a martingale w.r.t. filtration $(F_t)_{t\le 1}$ with $F_t \equiv F_{[nt]}$ and $\mathsf{E} z^n(\lambda) = 1$. Then, by martingale property

$$\mathsf{E} z_{\theta_c}^n(\lambda) = 1. \tag{1.84}$$

Pick $\lambda(s) \equiv r \geq 0$. Then, for the stopping time θ_c, in virtue of (1.84) and (1.57), we have:

$$\begin{aligned} 1 &\geq \mathsf{E}\,\mathrm{I}(W^n_{\theta_c} \geq c)\exp\left(n\left[\int_0^{\theta_c} r\,\mathrm{d}W^n_s - \int_0^{\theta_c} G_\mu(r)\,\mathrm{d}A^n_s\right]\right) \\ &\geq \mathsf{E}\,\mathrm{I}(W^n_{\theta_c} \geq c)\exp\left(n\left[rW^n_s - G_\mu(r)\int_0^1 \mathrm{d}A^n_s\right]\right) \\ &\geq \mathsf{E}\,\mathrm{I}(W^n_{\theta_c} \geq c)\exp\left(n[rc - G_\mu(r)]\right) \end{aligned}$$

and so, taking into account property (iii) of Lemma 1.1 in Section 1.1, we arrive at the following estimation:

$$n^{-1}\ln \mathsf{P}\,(W^n_{\theta_c} \geq c) \leq -\sup_{r\geq 0}\,[rc - G_\mu(r)] = -L(c).$$

Since, by property (iv) of Lemma 1.1 $L(c) \to \infty$ as $c \to \infty$ we get

$$\lim_{c\to\infty}\,\limsup_n n^{-1}\ln \mathsf{P}(W^n_{\theta_c} \geq c) \leq -\infty.$$

The first relation in (1.83) is proved. The second relation in (1.83) is proved similarly.

We next verify that

$$\lim_{\delta\to 0}\,\limsup_n\,\sup_{\tau\leq 1}\,n^{-1}\ln \mathsf{P}\left(\sup_{t\leq\delta}|W^n_{t+\tau} - W^n_\tau| \geq \eta\right) = -\infty. \tag{1.85}$$

For any stopping τ ($\tau \leq 1$) we define a process

$$x^n_t = W^n_{t+\tau} - W^n_\tau,$$

and put

$$\gamma_\eta = \inf\,(t:\ |x^n_t| \geq \eta) \wedge \delta.$$

Evidently, γ_η is a stopping time w.r.t. the filtration $(F_{t+\tau})_{t\leq 1}$ with $F_{\tau+t} \equiv F_{[n(\tau+t)]}$. Then

$$\begin{aligned} \left(\omega:\ \sup_{t\leq\delta}|W^n_{t+\tau} - W^n_\tau| \geq \eta\right) &= \left(\sup_{t\leq\delta}|x^n_t| \geq \eta\right) \\ &= \left(\sup_{t\leq\delta} x^n_t \geq \eta\right)\bigcup\left(\sup_{t\leq\delta} x^n_t \leq -\eta\right), \end{aligned}$$

and so

$$\mathsf{P}\left(\sup_{t\leq\delta}|W^n_{t+\tau} - W^n_\tau| \geq \eta\right) \leq 2\max\left(\mathsf{P}\,(x^n_{\gamma_\eta} \geq \eta), \mathsf{P}\,(x^n_{\gamma_\eta} \leq -\eta)\right).$$

Hence, (1.85) will be proved, once we show that

$$\begin{aligned} &\lim_{\delta\to 0}\,\limsup_n n^{-1}\ln\sup_{\tau\leq 1}\,\mathsf{P}(x^n_\gamma \geq \eta) \\ &\qquad = \lim_{\delta\to 0}\,\limsup_n n^{-1}\ln\sup_{\tau\leq 1}\,\mathsf{P}\,(x^n_\gamma \leq -\eta) = -\infty. \end{aligned} \tag{1.86}$$

Since $z^n_{t+\tau}(\lambda)$ is a martingale w.r.t. the filtration $(F_{t+\tau})$, $\tau \geq 0$, then

$$\frac{\mathsf{E}\,(z^n_{\tau+\gamma}(\lambda)|F_\tau)}{z^n_\tau(\lambda)} = 1.$$

The last equality (with $\lambda(s) \equiv r \geq 0$) implies that

$$\begin{aligned} 1 \;&\geq\; \mathsf{E}\,\mathrm{I}(x^n_\gamma \geq \eta)\exp\left(n\left[\int_\tau^{\tau+\gamma} r\,\mathrm{d}W^n_s - \int_\tau^{\tau+\gamma} G_\mu(r)\,\mathrm{d}A^n_s\right]\right) \\ &\geq\; \mathsf{E}\,\mathrm{I}(x^n_\gamma \geq \eta)\exp\left(n\left[r(W^n_{\tau+\gamma} - W^n_\tau) - \int_\tau^{\tau+\delta} G_\mu(r)\,\mathrm{d}A^n_s\right]\right) \\ &\geq\; \mathsf{E}\,\mathrm{I}(x^n_\gamma \geq \eta)\exp\left(n\left[r\eta - G_\mu(r)\frac{[n\delta]}{n}\right]\right). \end{aligned}$$

Denote $\delta_n = [n\delta]/n$ and observe that $\delta_n \to \delta$ as $n \to \infty$. Then, taking into account the property (iii) of Lemma 1.1, we get

$$n^{-1}\ln\sup_{\tau\leq 1}\mathsf{P}\,(x^n_\gamma \geq \eta) \leq -\sup_{r\geq 0}\,[r\eta - \delta_n G_\mu(r)] = -L\left(\frac{\eta}{\delta_n}\right).$$

Hence, by property (iv) of Lemma 1.1

$$\lim_{\delta\to 0}\,\limsup_n\, n^{-1}\ln\sup_{\tau\leq 1}\mathsf{P}\,(x^n_\gamma \geq \eta) = -\infty.$$

The second relation in (1.86) is proved similarly.

Proof of Theorem 1.13. We proved that the sequence X^n is C-exponentially tight. In Theorem 1.14 we obtained the lower bound for continuously differentiable functions φ. The passage to the lower bound with absolutely continuous φ is fulfilled by appropriate approximation (see Theorem 2.2 in Chapter 2).

In Theorem 1.15 we proved the upper bound with function $\tilde{I}(\varphi)$. It remains to show that $\tilde{I} = I$. The reader is referred to Theorem 2.3 in Chapter 2 for the proof. Theorem 1.13 is proved.

1.6. APPLICATIONS TO RECURSIVE ESTIMATION AND CONTROL THEORY

The Robbins-Monro process

Let $f(x)$ be the negative gradient of real valued function $V(x)$. The Robbins–Monro procedure is a way of stochastically solving the equation

$$f(x) = 0,$$

where to each value x there corresponds a random variable $Y = Y(x)$ with distribution $\mathsf{P}(Y(x) \leq y) = H(y|x)$ such that

$$f(x) = \int_R y\,\mathrm{d}H(y|x)$$

is the expectation of Y for given x. The Robbins–Monro procedure for finding the root of $f(x)$ is defined as (see, for example, Kushner and Clark (1978))

$$x(k) = x(k-1) + \gamma_k y(k),$$

where $x(k)$ is a parameter setting in the kth run; $y(k)$ is a random variable (observation) whose distribution function for given $x(1), \dots, x(k-1)$, $y(1), \dots, y(k-1)$ is $H(y|x(k-1))$; the sequence (γ_k), $k \geq 1$ assumed to be a sequence of positive scalars.

When it is assumed that estimates $x(k)$ are desired to converge to the optimal value x_0, $f(x_0)$, the 'noise' can only be rejected by paying less and less attention to the noisy observations, i.e., by letting $\gamma_k \to 0$, as $k \to \infty$.

In tracking problems, when a time varying parameter (parameters) is to be tracked using the Robbins–Monro algorithm, γ_k usually tends to small, non-zero value, the size of which depends on what is known about the variability of the tracked parameters and about the noise characteristics. Throughout this book we suppose that γ is small positive scalar. The methods of analysis are similar in both cases.

In Chapter 2 we consider the following general problem (for the sake of notation simplicity we discuss only the one-dimensional case).

Let $\xi = (\xi_k)_{k\geq 1}$ be a sequence of i.i.d. random variables with a common distribution μ, $\mathsf{E}\xi_1 = 0$, that are defined on a probability space (Ω, F, P) and take value in R. Consider a process

$$X_k^n - X_{k-1}^n = n^{-1}(b_k(X^{n,k-1}) + \sigma_k(X^{n,k-1})\xi_k), \quad X_0 = x, \quad k \geq 1,$$

where

$$b_k(X^{n,k-1}) = b_k(X_0^n, X_1^n, \dots, X_{k-1}^n),$$
$$\sigma_k(X^{n,k-1}) = \sigma_k(X_0^n, X_1^n, \dots, X_{k-1}^n),$$

with some b_k: $R^k \to R$ and σ_k: $R^k \to R$.

To put the problem in a continuous time context we introduce the family of processes

$$W^n = (W_t^n)_{0\leq t\leq 1}, \qquad W_t^n = n^{-1}\sum_{k=1}^{[nt]} \xi_k$$

and the sequence $X^n = (X_t^n)_{0\leq t\leq 1}$, $n \geq 1$, with

$$X_t^n = x + \int_0^t b^n\left(X_{[0,s)}^n\right)\, \mathrm{d}A_s^n + \int_0^1 \sigma^n\left(X_{[0,s)}^n\right)\, \mathrm{d}W_s^n,$$

where

$$A_t^n = \frac{[nt]}{n}, \qquad t_k = k/n, \qquad k \geq 1;$$

for each $X = (X_t)_{0\leq t\leq 1}$ in $D[0,1]$ b^n: $D[0,1] \to R$ and σ^n: $D[0,1] \to R$ are defined as

$$b^n(X_{[0,t)}) = b_{[nt]}(X_0, X_{t_1}, \dots, X_{t_k}, \dots, X_{[nt]-1}),$$

$$\sigma^n(X_{[0,t)}) = \sigma_{[nt]}(X_0, X_{t_1}, \dots, X_{t_k}, \dots, X_{[nt]-1}),$$

and

$$\int_0^t b^n\left(X^n_{[0,s)}\right) \mathrm{d}A^n_s = n^{-1}\sum_{k=1}^{[nt]} b_k(X^{n,k-1}),$$

$$\int_0^1 \sigma^n\left(X^n_{[0,s)}\right) \mathrm{d}W^n_s = n^{-1}\sum_{k=1}^{[nt]} \sigma_k(X^{n,k-1})\xi_k.$$

Assume that there exists a continuous mapping b: $C[0,1] \to R$ such that for every $X \in C[0,1]$

$$\lim_n \int_0^1 \left|b^n\left(X_{[0,s)}\right) - b\left(X_{[0,s)}\right)\right| \mathrm{d}A^n_s = 0.$$

Under reasonable condition (X^n) converges as $n \to \infty$ to the trajectory of the ordinary differential equation

$$X_t = x + \int_0^t b\left(X_{[0,s)}\right) \mathrm{d}s, \qquad X_0 = x.$$

The problem in which we are interested is that of determining an asymptotic behaviour of the probability

$$\mathsf{P}\left(\sup_{t\leq 1} |X^n_t - \varphi_t| \leq \delta\right),$$

i.e., what is the probability that the trajectory X^n_t be close in the time interval $[0,1]$ to a preassigned function φ_t? Thus, we come to the large deviation problem in a function space.

Since large deviations exploit more of the structure of the algorithm than 'local' limits of the type of suitably normalized X^n do, this approach often provides more comprehensive information and is important in studying the asymptotic properties of X^n (see Dupuis and Kushner (1987, 1989)).

In Chapter 2 we give the proof of the large deviation property for this scheme, based on the concept of stochastic exponent. In Chapter 3 we consider an extension of the problem to the case when the 'noise' (ξ_n) is a stationary process which admits the Wold-type decomposition.

An identification problem

An important class of applications concerns recursive procedures for estimating the parameters of linear system.

Let (ξ_n), $n \geq 1$, be a sequence of i.i.d. random variables and $\theta^* = (a^*_1, \dots, a^*_k, c^*_1, \dots, c^*_l)$ be a set of given unknown scalars. Define a sequence (y_n), $n \geq 1$, by the input output system equations

$$y_n + \sum_{i=1}^k a^*_i y_{n-i} = \sum_{i=1}^l c^*_i \xi_{n-i} + \xi_n$$

or in the operator form,

$$A^*(q^{-1})y_n = C^*(q^{-1})\xi_n,$$

where $A^*(q^{-1})$ and $C^*(q^{-1})$ are polynomials in the backward shift operator q^{-1}. Our goal is to estimate the parameters θ^*.

Let $\theta_n = (a_1(n), \dots, a_k(n), c_1(n), \dots, c_l(n))$ be a vector of parameter estimates based on the data $y_1, \dots, y_n$. We specify the recursion by a formula for its one step prediction error or innovations sequence (ε_n), $n \geq 1$,

$$C_n(q^{-1})\varepsilon_n = A_n(q^{-1})y_n,$$

where $A(q^{-1})$ and $C(q^{-1})$ are polynomials in the backward shift operator q^{-1} with the parameters $(c_1(n), \dots, c_l(n))$ and $(a_1(n), \dots, a_k(n))$ accordingly. Define

$$\varphi_n^T = (y_{n-1}, \dots, y_{n-k}, -\varepsilon_{n-1}, \dots, -\varepsilon_{n-l})$$

and the recursion

$$\theta_{n+1} = \theta_n + \gamma_n \varphi_n \varepsilon_n$$

with the sequence (γ_n), $n \geq 1$, assumed to be a sequence of positive scalars or small non-zero scalar.

Assume that the pair of polynomials $A(z)\frac{C^*(z)}{A^*(z)}$ and $C_n(z)$ (z is an arbitrary complex variable, replacing q^{-1}) have all their zeros outside the unit circle. Then (ε_n) is a stationary process with a rational spectrum and hence could be represented as a component of some p-dimensional ($p = k + l$) Markov process Y_n (see, for example, Liptser and Shiryaev (1977)):

$$Y_{n+1} = A(\theta_n)Y_n + B(\theta_n)W_n, \qquad n \geq 0,$$

where $A(\theta)$ and $B(\theta)$ are matrices of appropriate dimensions and with components, expressed in terms of the parameters θ^* and θ; W_n is a sequence of independent vectors.

In fact, we can consider the general recursive procedure

$$\theta_{n+1} = \theta_n + \gamma_n f(\theta_n, Y_{n+1}), \tag{1.87}$$

$$Y_{n+1} = Y_n + a_n(\theta_n, Y_n) + b_n(\theta_n, Y_n)W_n, \tag{1.88}$$

$n \geq 0$, where $\theta_n \in E^p$, $Y_n \in E^d$; $a_n(\theta, y)$ is a d-dimensional vector; $b_n(\theta, y)$ is a $m \times d$-matrix; W_n, $n \geq 0$, is a sequence of independent vectors, which takes value in E^m.

Note that in this scheme the observations are in general functions of the previous estimates $\theta_{(\cdot)}$ and of a sequence of random variables (W_n). This is the case, e.g., for adaptive systems, when the input signal is determined on the basis of previous estimates.

Suppose we have a linear, stochastic, discrete-time system, governed by a linear output feedback law, which at time n is determined by θ_{n-1}. Then the behaviour of this over-all system can be described by (1.87), with Y_n consisting of lagged inputs and outputs. Therefore the algorithms (1.87) and (1.88) can be understood as archetypical for adaptive control of linear system (see Ljung (1977)).

In Chapter 6 we sketch the proof of the averaging principle and in Chapter 11 we prove the large deviation principle for this algorithm.

Chapter 2

Large deviations for the non-Markovian recursive scheme with additive 'white noise'

In this chapter we continue our program and consider the large deviation property for a certain recursive scheme.

Setting
Let $\xi = (\xi_k)_{k\geq 1}$ be a sequence of i.i.d. random variables with a common distribution μ, $\mathsf{E}\xi_1 = 0$, that are defined on a probability space (Ω, F, P) and that take values in R. Define $(F_k)_{k\geq 1}$ a family of σ-fields in Ω, generated by $\xi_1, \ldots, \xi_k$, $F_k = \sigma(\xi_1, \ldots, \xi_k)$. Assume that *the moment generating function* (Cramér's condition)

$$\mathsf{E}\exp(\lambda\xi_1)$$

is finite for all $\lambda \in R$ and *the logarithmic moment generating function*

$$G_\mu(\lambda) = \ln \mathsf{E}\exp(\lambda\xi_1),$$

satisfies the regularity condition

$$\lim_{|\lambda|\to\infty} \frac{G_\mu(\lambda)}{|\lambda|} = \infty. \tag{R}$$

Consider a process

$$X_k - X_{k-1} = \gamma(b(X_{k-1}) + \xi_k), \qquad X_0 = x, \qquad k \geq 1.$$

with $b\colon R \to R$ and $\gamma = 1/n$. Introduce a family of processes $X^n = (X_t^n)_{0\leq t\leq 1}$ with

$$X_t^n = x + \int_0^t b(X_{s-}^n)\, \mathrm{d}A_s^n + W_t^n,$$

being a 'continuous time' version of the recursive procedure, where, as in Section 1.6,

$$A_t^n = [nt]/n,$$

$$W_t^n = n^{-1}\sum_{k=1}^{[nt]} \xi_k,$$

and

$$\int_0^t b(X_{s-}^n)\, \mathrm{d}A_s^n = n^{-1}\sum_{k=1}^{[nt]} b(X_{k-1})$$

with X_{s-} being the left-hand limit of X_s.

It is well known (see, e.g., Ljung (1977) and Chapter 6) that under natural assumption (X^n) converges as $n \to \infty$ to the trajectory of the ordinary differential equation

$$\dot{x}(t) = b(x_t), \qquad x(0) = x.$$

Hence, we may ask about an asymptotic behaviour of the probability

$$\mathsf{P}\left(\sup_{t \leq 1} (X_t^n - \varphi_t) < \delta\right),$$

for $\varphi \in C[0,1]$.

In fact, we are able to carry out this program even for more sophisticated scheme. Consider a process

$$X_k^n - X_{k-1}^n = n^{-1}(b_k(X^{n,k-1}) + \sigma_k(X^{n,k-1})\xi_k), \qquad X_0 = x, \qquad k \geq 1, \tag{2.1}$$

where

$$b_k(X^{n,k-1}) = b_k(X_0^n, X_1^n, \dots, X_{k-1}^n),$$
$$\sigma_k(X^{n,k-1}) = \sigma_k(X_0^n, X_1^n, \dots, X_{k-1}^n).$$

Denote

$$\Delta = 1/n, \qquad t_k = k\Delta, \qquad k \geq 1,$$

and for each $X = (X_t)_{0 \leq t \leq 1}$, define $b^n\colon D([0,1]) \to R$, and $\sigma^n\colon D([0,1]) \to R$ such that

$$\begin{aligned} b^n(X_{[0,t)}) &= b_{[nt]}(X_0, X_{t_1}, \dots, X_{t_k}, \dots, X_{[nt]-1}), \\ \sigma^n(X_{[0,t)}) &= \sigma_{[nt]}(X_0, X_{t_1}, \dots, X_{t_k}, \dots, X_{[nt]-1}). \end{aligned} \tag{2.2}$$

We will study large deviations for the sequence $X^n = (X_t^n)_{0 \leq t \leq 1}$, $n \geq 1$, with

$$X_t^n = x + \int_0^t b^n(X_{[0,s)}^n)\, dA_s^n + \int_0^t \sigma^n(X_{[0,s)}^n)\, dW_s^n, \tag{2.3}$$

where

$$\int_0^t b^n(X_{[0,s)}^n)\, dA_s^n = n^{-1} \sum_{k=1}^{[nt]} b_k(X^{n,k-1}),$$

$$\int_0^t \sigma^n(X_{[0,s)}^n)\, dW_s^n = n^{-1} \sum_{k=1}^{[nt]} \sigma_k(X^{n,k-1})\xi_k.$$

Assume that $b^n(X)$ and $\sigma^n(X)$ satisfy the following conditions:

(i) For all $X \in D[0,1]$

$$\sup_n |b^n(X)| \leq K,$$
$$\sup_n |\sigma^n(X)| \leq K$$

with some $K > 0$.

(ii) For every $X, Y \in D[0,1]$,

$$\sup_n |b^n(X_{[0,t)}) - b^n(Y_{[0,t)})| \le K \sup_{s<t} |X_s - Y_s|,$$

$$\sup_n |\sigma^n(X_{[0,t)}) - \sigma^n(Y_{[0,t)})| \le K \sup_{s<t} |X_s - Y_s|.$$

(iii) There exists a continuous mapping b: $C[0,1] \to R$ such that for every $X \in C[0,1]$

$$\lim_n \int_0^1 |b^n(X_{[0,s)}) - b(X_{[0,s)})| \, \mathrm{d}A_s^n = 0,$$

and continuous mapping σ: $C[0,1] \to R$, $\sigma^2 > 0$ such that for every $X \in C[0,1]$

$$\lim_n \int_0^1 |\sigma^n(X_{[0,s)}) - \sigma(X_{[0,s)})| \, \mathrm{d}A_s^n = 0,$$

where

$$\int_0^t b^n(X_{[0,s)}) \, \mathrm{d}A_s^n = n^{-1} \sum_{k=1}^{[nt]} b_k(X^{k-1}),$$

$$\int_0^t \sigma^n(X_{[0,s)}) \, \mathrm{d}A_s^n = n^{-1} \sum_{k=1}^{[nt]} \sigma_k(X^{k-1}).$$

Now we state the main result.

Theorem 2.1. Let Cramér's condition, the regularity condition (R) and conditions (i)–(iii) hold. Then the sequence X^n, $n \ge 1$, obeys the large deviation principle in $D[0,1]$ with the rate function

$$I(\varphi) = \begin{cases} \int_0^1 L(\sigma^{-1}(\varphi_{[0,s]})[\dot\varphi - b(\varphi_{[0,s]})]) \, dt, & \varphi \text{ is absolutely continuous, } \varphi_0 = x, \\ \infty, & \text{otherwise}. \end{cases}$$

with

$$L(x) = \sup_{\lambda \in R} [\lambda x - G_\mu(\lambda)].$$

Remark 2.1. Assume that (ξ_k) are Gaussian with $\mathsf{E}\xi^2 = 1$ then

$$G(\lambda) = \ln \mathsf{E} \exp(\lambda \xi_1) = \lambda^2/2$$

and

$$I(\varphi) = \begin{cases} \int_0^1 (\sigma^{-1}(\varphi_{[0,s]})[\dot\varphi - b(\varphi_{[0,s]})])^2 \, dt, & \varphi \text{ is absolutely continuous, } \varphi_0 = x, \\ \infty, & \text{otherwise}. \end{cases}$$

Remark 2.2. The result remains valid if condition (i) is replaced by condition of linear growth.

Stochastic exponent and cummulant. Transformation of measure

There are several ways in which one can go about studying large deviations for the sequence X^n. We choose one which emphasizes the role of the stochastic moment generating function and the stochastic exponential.

Denote $\Delta X_k^n = X_k^n - X_{k-1}^n$. Given numerical sequence $\lambda = (\lambda_k)$, $\lambda_k \in R$, define *the stochastic moment generating function* (the stochastic exponent)

$$\mathcal{E}_k(\lambda, X^{n,k-1}) = \prod_{j=1}^{k} \mathsf{E}\,(\exp(\lambda_j \Delta X_j^n) | F_{j-1})$$

and *the stochastic logarithmic moment generating function* (the stochastic cummulant)

$$\ln \mathcal{E}_k(\lambda, X^{n,k-1}) = \sum_{j=1}^{k} \ln \mathsf{E}\,(\exp(\lambda_j \Delta X_j^n) | F_{j-1}).$$

By (2.1), taking into account that $b_j(X^{n,j-1})$ is F_{j-1}-measurable, we write the cummulant in the following form:

$$\begin{aligned}
\ln \mathcal{E}_k(n\lambda, X^{n,k-1}) &= \sum_{j=1}^{k} \ln \mathsf{E}\,(\exp[\lambda_j b_j(X^{n,j-1}) + \lambda_j \sigma_j(X^{n,j-1})\xi_j] | F_{j-1}) \\
&= \sum_{j=1}^{k} \ln[e^{\lambda_j b_j(X^{n,j-1})} \mathsf{E}\,(e^{\lambda_j \sigma_j(X^{n,j-1})\xi_j} | F_{j-1})] \\
&= \sum_{j=1}^{k} \lambda_j b_j(X^{n,j-1}) + \sum_{j=1}^{k} \ln \mathsf{E}\,(e^{\lambda_j \sigma_j(X^{n,j-1})\xi_j} | F_{j-1}).
\end{aligned}$$

Since ξ is a sequence of i.i.d variables and $\sigma_j(X^{n,j-1})$ is F_{j-1}-measurable we have

$$\ln \mathcal{E}_k(n\lambda, X^{n,k-1}) = \sum_{j=1}^{k} \lambda_j b_j(X^{n,j-1}) + \sum_{j=1}^{k} G_\mu(\lambda_j \sigma_j(X^{n,j-1})),$$

where

$$\ln \mathsf{E}\,(e^{\lambda_j \sigma_j(X^{n,j-1})\xi_j} | F_{j-1}) = G_\mu(\lambda_j \sigma_j(X^{n,j-1})).$$

Just as in the previous Section 1.5 the key to proving the lower bound is the use of Cramér's idea for moving the mean of the distribution of X_t^n. To this end, we introduce the process (*the stochastic exponential*)

$$\begin{aligned}
z_m^n(\lambda) &= \prod_{k=1}^{m} \frac{\exp(\lambda_k n \Delta X_k^n)}{\mathsf{E}\,(\exp(\lambda_k n \Delta X_k^n) | F_{k-1})} \\
&= \exp\left(n \sum_{k=1}^{m} \lambda_k \Delta X_k^n - \ln \mathcal{E}_m(n\lambda, X^{n,m-1})\right),
\end{aligned} \tag{2.4}$$

and denote $z^n(\lambda) = z_n^n(\lambda)$. It is easy to see that $(z_k^n(\lambda), F_k)$ is a martingale. Indeed,

$$\begin{aligned}
\mathsf{E}\,(z_m^n(\lambda) | F_{m-1}) &= \mathsf{E}\left(\prod_{k=1}^{m} \frac{\exp(\lambda_k n \Delta X_k^n)}{\mathsf{E}\,(\exp(\lambda_k n \Delta X_k^n) | F_{k-1})} | F_{m-1}\right) \\
&= z_{m-1}^n(\lambda)\, \mathsf{E}\left(\frac{\exp(\lambda_m n \Delta X_m^n)}{\mathsf{E}\,(\exp(\lambda_m n \Delta X_m^n) | F_{m-1})} | F_{m-1}\right) \\
&= z_{m-1}^n(\lambda).
\end{aligned} \tag{2.5}$$

Evidently, $\mathsf{E}z^n(\lambda) = 1$ and so we can define a new probability measure $Q^{n,\lambda}$ by

$$dQ^{n,\lambda} = z^n(\lambda)\, d\mathsf{P}. \tag{2.6}$$

The definition of $z^n(\lambda)$ implies that $z^n(\lambda) > 0$ P-a.s. So the measures P and $Q^{n,\lambda}$ are equivalent, and

$$d\mathsf{P} = (z^n(\lambda))^{-1}\, dQ^{n,\lambda}. \tag{2.7}$$

To put the problem in an infinite dimensional context we need some notations. For a continuous function $\lambda(s)$, $0 \le s \le 1$, with the property $\lambda(k\Delta) = \lambda_k$, $k = 0, 1, \ldots$, and any $Y \in D[0,1]$ with $Y_k = Y(k\Delta)$ define

$$\ln \mathcal{E}_t(n\lambda, Y) = n \int_0^t \lambda(s)\, b^n(Y_{[0,s)})\, dA_s^n + n \int_0^t G_\mu(\lambda(s)\,\sigma^n(Y_{[0,s)}))\, dA_s^n, \tag{2.8}$$

with

$$\begin{aligned} \int_0^t \lambda(s)\, b^n(Y_{[0,s)})\, dA_s^n &= n^{-1} \sum_{k=1}^{[nt]} \lambda_k b_k(Y^{k-1}), \\ \int_0^t G_\mu(\lambda(s)\,\sigma^n(Y_{[0,s)}))\, dA_s^n &= n^{-1} \sum_{k=1}^{[nt]} G_\mu(\lambda_k \sigma_k(Y^{k-1})). \end{aligned} \tag{2.9}$$

Denote

$$\int_0^t \lambda(s)\, dX_s^n = \sum_{k=1}^{[nt]} \lambda_k \Delta X_k^n,$$

and observe that, in virtue of (2.3),

$$\int_0^t \lambda(s)\, dX_s^n = \int_0^t \lambda(s)\, b^n(X_{[0,s)}^n)\, dA_s^n + \int_0^t \lambda(s)\,\sigma^n(X_{[0,s)}^n)\, dW_s^n.$$

Taking into account this notation and (2.8), we define

$$\begin{aligned} z_t^n(\lambda) &= \exp\left(n \int_0^t \lambda(s)\, dX_s^n - \ln \mathcal{E}_t(n\lambda, Y)\right) \\ &= \exp\left(n\left[\int_0^t \lambda(s)\, dX_s^n - \int_0^t \lambda(s)\, b^n(X_{[0,s)}^n)\, dA_s^n \right.\right. \\ &\qquad \left.\left. - \int_0^t G_\mu(\lambda(s)\,\sigma^n(X_{[0,s)}^n))\, dA_s^n\right]\right) \\ &= \exp\left(n\left(\int_0^t \lambda(s)\,\sigma^n(X_{[0,s)}^n)\, dW_s^n - \int_0^t G_\mu(\lambda(s)\,\sigma^n(X_{[0,s)}^n))\, dA_s^n\right)\right), \end{aligned} \tag{2.10}$$

$0 \le t \le 1$, and put $z_1^n(\lambda) = z^n(\lambda)$. Denote $F_t = F_{[nt]}$ and observe that $z_t^n(\lambda)$, $0 \le t \le 1$, is a martingale w.r.t. the filtration (F_t), $0 \le t \le 1$.

Decomposition X_t^n w.r.t $Q^{n,\lambda}$

Following the outline of Section 1.5, we now introduce a decomposition X_t^n w.r.t the measure $Q^{n,\lambda}$, defined by (2.6) and (2.10). Unlike the mentioned case, ξ is not the sequence of independent variables w.r.t the measure $Q^{n,\lambda}$. Nevertheless, we can represent $\int_0^1 \sigma^n(X^n_{[0,s)})\, dW^n_s$ as a sum of a square integrable martingale (w.r.t $Q^{n,\lambda}$) and a compensator ($\mathsf{E}^{n,\lambda}$ denotes an expectation w.r.t. $Q^{n,\lambda}$):

$$\begin{aligned}
\int_0^t \sigma^n(X^n_{[0,s)})\, dW^n_s &= n^{-1}\sum_{k=1}^{[nt]} \sigma_k(X^{n,k-1})\xi_k \\
&= n^{-1}\sum_{k=1}^{[nt]} [\sigma_k(X^{n,k-1})\xi_k - \mathsf{E}^{n,\lambda}(\sigma_k(X^{n,k-1})\xi_k | F_{k-1})] \\
&\quad + n^{-1}\sum_{k=1}^{[nt]} \mathsf{E}^{n,\lambda}(\sigma_k(X^{n,k-1})\xi_k | F_{k-1}) \\
&= n^{-1}\sum_{k=1}^{[nt]} \sigma_k(X^{n,k-1})[\xi_k - \mathsf{E}^{n,\lambda}(\xi_k | F_{k-1})] \\
&\quad + n^{-1}\sum_{k=1}^{[nt]} \sigma_k(X^{n,k-1})\, \mathsf{E}^{n,\lambda}(\xi_k | F_{k-1}) \\
&= M_t^n(\lambda) + B_t^n(\lambda), \qquad (2.11)
\end{aligned}$$

with

$$M_t^n(\lambda) = n^{-1}\sum_{k=1}^{[nt]} \sigma_k(X^{n,k-1})[\xi_k - \mathsf{E}^{n,\lambda}(\xi_k | F_{k-1})]$$

and

$$B_t^n(\lambda) = n^{-1}\sum_{k=1}^{[nt]} \sigma_k(X^{n,k-1})\, \mathsf{E}^{n,\lambda}(\xi_k | F_{k-1}).$$

Evidently, under condition (i), $(M_t^n(\lambda), F_t^n)$ is a square integrable martingale w.r.t $Q^{n,\lambda}$.

We next calculate the conditional expectation $\mathsf{E}^{n,\lambda}(\xi_k | F_{k-1})$, $k \le n$. Denote (see (2.4))

$$\alpha_k(\lambda) = \frac{\exp(\lambda_k n \Delta X_k^n)}{\mathsf{E}(\exp(\lambda_k n \Delta X_k^n) | F_{k-1})} = \frac{\exp(\lambda_k \sigma_k(X^{n,k-1})\xi_k)}{\mathsf{E}(\exp(\lambda_k \sigma_k(X^{n,k-1})\xi_k) | F_{k-1})}.$$

It is well known (see, e.g., Liptser and Shiryaev (1989)) that $\mathsf{E}^{n,\lambda}(\xi_k | F_{k-1})$ can be expressed w.r.t. measure P by the following formula:

$$\begin{aligned}
\mathsf{E}^{n,\lambda}(\xi_k | F_{k-1}) &= \mathsf{E}(\alpha_k(\lambda)\xi_k | F_{k-1}) \\
&= \frac{\mathsf{E}(\xi_k \exp(\lambda_k \sigma_k(X^{n,k-1})\xi_k) | F_{k-1})}{\mathsf{E}(\exp(\lambda_k \sigma_k(X^{n,k-1})\xi_k) | F_{k-1})}. \qquad (2.12)
\end{aligned}$$

At the same time

$$\begin{aligned}
\frac{dG_\mu(\beta)}{d\beta}\Big|_{\beta = \lambda_k \sigma_k(X^{n,k-1})} &= \frac{\mathsf{E}(\xi_k \exp(\lambda_k \sigma_k(X^{n,k-1})\xi_k) | F_{k-1})}{\mathsf{E}(\exp(\lambda_k \sigma_k(X^{n,k-1})\xi_k) | F_{k-1})} \\
&= G'_\mu(\lambda_k \sigma_k(X^{n,k-1})), \qquad (2.13)
\end{aligned}$$

and hence

$$\mathsf{E}^{n,\lambda}(\xi_k|F_{k-1}) = G'_\mu(\lambda_k\sigma_k(X^{n,k-1})). \tag{2.14}$$

Combining (2.11), (2.14) and (2.3), we arrive at the following representation

$$\begin{aligned} X_t^n &= x + \int_0^t b^n(X_{[0,s)}^n)\, \mathrm{d}A_s^n + M_t^n(\lambda) \\ &+ \int_0^t \sigma^n(X_{[0,s)}^n)\, G'_\mu(\lambda(s)\,\sigma^n(X_{[0,s)}^n))\, \mathrm{d}A_s^n, \end{aligned} \tag{2.15}$$

with

$$\begin{aligned} &\int_0^t \sigma^n(X_{[0,s)}^n)\, G'_\mu(\lambda(s)\,\sigma^n(X_{[0,s)}^n))\, \mathrm{d}A_s^n \\ &\qquad = n^{-1}\sum_{k=1}^{[nt]} \sigma_k(X_n^{k-1})\, G'_\mu(\lambda_k\sigma_k(X_n^{k-1})) = B_t^n(\lambda). \end{aligned} \tag{2.16}$$

Ergodic property X_t^n w.r.t $Q^{n,\lambda}$

Consider a continuously differentiable function $\varphi(s)$, $0 \le s \le 1$, with $\varphi(0) = X_0^n = x$. We will show in this paragraph that with appropriate $\lambda(s)$

$$\lim_n Q^{n,\lambda}\left(\sup_{t\le 1} |X^n(t) - \varphi(t)| > \delta\right) = 0, \tag{2.17}$$

for every $\delta > 0$. It means in the present setting that we move 'the centre' of the distribution of X_t^n asymptotically to $\varphi(t)$. We use the decomposition (2.15) to get on the guess how to choose $\lambda(s)$.

Consider

$$\begin{aligned} X_t^n - \varphi(t) &= X_t^n - \int_0^t \dot\varphi(s)\, \mathrm{d}A_s^n + \int_0^t \dot\varphi(s)\, \mathrm{d}(A_s^n - s) \\ &= X_t^n - \int_0^t \dot\varphi(s)\, \mathrm{d}A_s^n + J_1^n(t), \end{aligned}$$

with

$$J_1^n(t) = \int_0^t \dot\varphi(s)\, \mathrm{d}(A_s^n - s).$$

Taking into account (2.15), we write:

$$\begin{aligned} X_t^n - \varphi(t) &= M_t^n(\lambda) - \int_0^t [\dot\varphi(s) - b^n(X_{[0,s)}^n) \\ &\quad -\sigma^n(X_{[0,s)}^n)\, G'_\mu(\lambda(s)\,\sigma^n(X_{[0,s)}^n))]\, \mathrm{d}A_s^n + J_1^n(t). \end{aligned} \tag{2.18}$$

Taking a hint from this decomposition, one is temped to guess for choosing $\lambda(s) = \lambda^0(s)$ such that

$$\dot{\varphi}(s) + b(\varphi_{[0,s]}) - \sigma(\varphi_{[0,s]})\, G'_\mu(\lambda^0(s)\,\sigma(\varphi_{[0,s]})) = 0. \tag{2.19}$$

Indeed, denote

$$J^n_{2,3}(t) = \int_0^t (b^n(X^n_{[0,s)}) - b(\varphi_{[0,s]}))\, \mathrm{d}A^n_s \tag{2.20}$$

and

$$\begin{aligned} J^n_{4,5}(t) \;&=\; \int_0^t [\sigma^n(X^n_{[0,s)})\, G'_\mu(\lambda(s)\,\sigma^n(X^n_{[0,s)})) \\ &\quad -\sigma(\varphi_{[0,s]})\, G'_\mu(\lambda(s)\,\sigma(\varphi_{[0,s]}))]\, \mathrm{d}A^n_s, \end{aligned} \tag{2.21}$$

and, taking into account (2.19), rewrite (2.18) in the following form:

$$\begin{aligned} X^n_t - \varphi(t) \;&=\; M^n_t(\lambda) - \int_0^t [\dot{\varphi}(s) + b^n(\varphi_{[0,s]}) \\ &\quad -\sigma(\varphi_{[0,s]})\, G'_\mu(\lambda(s)\,\sigma(\varphi_{[0,s]}))]\, \mathrm{d}A^n_s + J^n_1(t) + J^n_{2,3}(t) + J^n_{4,5}(t) \\ &=\; M^n_t(\lambda) + J^n_1(t) + J^n_{2,3}(t) + J^n_{4,5}(t). \end{aligned} \tag{2.22}$$

We have seen in Section 1.5 that $J^n_1 \to 0$ as $n \to \infty$. At the same time, we may expect that under our conditions $J^n_{2,3}, J^n_{4,5} \to 0$ as $n \to \infty$. Just as in Section 1.5 we will then be able to use the martingale property for proving (2.17).

Turning back to (2.19), we note that $\lambda^0(s)$ is the solution of the equation

$$\frac{\mathrm{d}}{\mathrm{d}\lambda}(\lambda(\dot{\varphi}(s) - b(\varphi_{[0,s]})) - G_\mu(\lambda\sigma(\varphi_{[0,s]}))) = 0. \tag{2.23}$$

Under our conditions $\lambda^0(s)$, $0 \le s \le 1$, is a continuous bounded function and

$$\begin{aligned} &\sup_\lambda (\lambda[\dot{\varphi}(s) - b(\varphi_{[0,s]})] - G_\mu(\lambda\sigma(\varphi_{[0,s]}))) \\ &\qquad = L(\sigma^{-1}(\varphi_{[0,s]})[\dot{\varphi}(s) - b(\varphi_{[0,s]})]), \end{aligned} \tag{2.24}$$

with

$$L(y) = \sup_\beta (\beta y - G_\mu(\beta)).$$

We continue our rigorous analysis with the lemma which shows that the heuristic reasoning above is consistent.

Lemma 2.1. Under conditions of Theorem 2.1

$$\lim_n Q^{n,\lambda}\left(\sup_{t\le 1} |X^n(t) - \varphi(t)| > \delta\right) = 0,$$

for every $\delta > 0$.

Proof. We first write $J^n_{2,3}(t)$, $J^n_{4,5}(t)$ (see (2.20) and (2.21)) in the following form:

$$\begin{aligned} J^n_{2,3}(t) &= \int_0^t (b^n(X^n_{[0,s)}) - b^n(\varphi_{[0,s]}))\, dA^n_s + \int_0^t (b^n(\varphi_{[0,s)}) - b(\varphi_{[0,s]}))\, dA^n_s \\ &= J^n_2(t) + J^n_3(t) \end{aligned} \tag{2.25}$$

$$\begin{aligned} J^n_{4,5}(t) &= \int_0^t [\sigma^n(X^n_{[0,s)})G'_\mu(\lambda^0(s)\sigma^n(X^n_{[0,s)})) - \sigma^n(\varphi_{[0,s]})G'_\mu(\lambda^0(s)\sigma^n(\varphi_{[0,s]}))]\, dA^n_s \\ &\quad + \int_0^t [\sigma^n(\varphi_{[0,s]})G'_\mu(\lambda^0(s)\sigma^n(\varphi_{[0,s]})) - \sigma(\varphi_{[0,s]})G'_\mu(\lambda^0(s)\sigma(\varphi_{[0,s]}))]\, dA^n_s \\ &= J^n_4(t) + J^n_5(t). \end{aligned} \tag{2.26}$$

Denote

$$\Delta^n_t = \sup_{s\le t} |X^n_s - \varphi(s)|$$

and observe that, by condition (ii),

$$|J^n_2(t)| \le \int_0^t \sup_{u\le s} |b^n(X^n_{[0,u)}) - b^n(\varphi_{[0,u]})|\, dA^n_s \le \text{const} \int_0^t \Delta^n_{s-}\, dA^n_s. \tag{2.27}$$

Next note that

$$\begin{aligned} |J^n_4(t)| &= \left|\int_0^t [\sigma^n(X^n_{[0,s)})G'_\mu(\lambda^0(s)\sigma^n(X^n_{[0,s)})) - \sigma^n(\varphi_{[0,s]})G'_\mu(\lambda^0(s)\sigma^n(\varphi_{[0,s]}))]\, dA^n_s\right| \\ &\le \left|\int_0^t [\sigma^n(X^n_{[0,s)})G'_\mu(\lambda^0(s)\sigma^n(X^n_{[0,s)})) - \sigma^n(\varphi_{[0,s]})G'_\mu(\lambda^0(s)\sigma^n(X^n_{[0,s)}))]\, dA^n_s\right| \\ &\quad + \left|\int_0^t [\sigma^n(\varphi_{[0,s]})G'_\mu(\lambda^0(s)\sigma^n(X^n_{[0,s)})) - \sigma^n(\varphi_{[0,s]})G'_\mu(\lambda^0(s)\sigma^n(\varphi_{[0,s]}))]\, dA^n_s\right|. \end{aligned}$$

Since $G_\mu(\lambda)$ is a smooth function, and, in view of conditions (i) and (ii), by repeating the preceding argument, we get

$$|J^n_4(t)| \le \text{const} \int_0^t \Delta^n_{s-}\, dA^n_s. \tag{2.28}$$

Combining (2.22) and (2.25)–(2.28), we arrive at the following inequality:

$$\begin{aligned} \Delta^n_t &= \sup_{s\le t} |M^n_s(\lambda) + J^n_1(s) + J^n_2(s) + J^n_3(s) + J^n_4(s) + J^n_5(s)| \\ &\le \sup_{t\le 1} |M^n_t(\lambda^0)| + \text{const} \int_0^t \Delta^n_{s-}\, dA^n_s \\ &\quad + \sup_{t\le 1} |J^n_1(t)| + \sup_{t\le 1} |J^n_3(t)| + \sup_{t\le 1} |J^n_5(t)|. \end{aligned}$$

We finally use a stochastic version of Gronwall–Bellman's formula to obtain

$$\Delta_t^n \leq \left(\sup_{t\leq 1} |M_t^n(\lambda^0)| + \sup_{t\leq 1} |J_1^n(t)| + \sup_{t\leq 1} |J_3^n(t)| + \sup_{t\leq 1} |J_5^n(t)|\right) \exp(K_1),$$

with some $K_1 < \infty$.

Hence, the proof will be complete, once we show that

$$\lim_n \sup_{t\leq 1} |J_1^n(t)| = 0, \tag{2.29}$$

$$\lim_n \sup_{t\leq 1} |J_3^n(t)| = 0, \tag{2.30}$$

$$\lim_n \sup_{t\leq 1} |J_5^n(t)| = 0, \tag{2.31}$$

and

$$\lim_n Q^{n,\lambda^0}\left(\sup_{t\leq 1} |M_t^n(\lambda^0)| \geq \delta\right) = 0,$$

for every $\delta > 0$.

We first observe that, by Lemma 1.6 in Section 1.5, (2.29) holds. Next note (cf. (2.25)) that

$$\begin{aligned} \sup_{t\leq 1} |J_3^n(t)| &= \sup_{t\leq 1} \left| \int_0^t [b^n(\varphi_{[0,s]}) - b(\varphi_{[0,s]})] \, dA_s^n \right| \\ &\leq \int_0^1 |b^n(\varphi_{[0,s]}) - b(\varphi_{[0,s]})| \, dA_s^n \end{aligned} \tag{2.32}$$

and, by condition (iii), (2.30) holds.

Turning to the proof of (2.31) we first obtain the following estimation:

$$\begin{aligned} |J_5^n(t)| &= \left| \int_0^t [\sigma^n(\varphi_{[0,s]}) G'_\mu(\lambda^0(s)\sigma^n(\varphi_{[0,s]})) - \sigma(\varphi_{[0,s]}) G'_\mu(\lambda^0(s)\sigma(\varphi_{[0,s]}))] \, dA_s^n \right| \\ &\leq \left| \int_0^t [\sigma^n(\varphi_{[0,s]}) G'_\mu(\lambda^0(s)\sigma^n(\varphi_{[0,s]})) - \sigma(\varphi_{[0,s]}) G'_\mu(\lambda^0(s)\sigma^n(\varphi_{[0,s]}))] \, dA_s^n \right| \\ &\quad + \left| \int_0^t [\sigma(\varphi_{[0,s]}) G'_\mu(\lambda^0(s)\sigma^n(\varphi_{[0,s]})) - \sigma(\varphi_{[0,s]}) G'_\mu(\lambda^0(s)\sigma(\varphi_{[0,s]}))] \, dA_s^n \right| \end{aligned}$$

Taking into account that $G_\mu(\lambda)$ is a smooth function, by condition (i), we obtain

$$\sup_{t\leq 1} |J_5^n(t)| \leq \text{const} \int_0^1 |\sigma^n(\varphi_{[0,s]}) - \sigma(\varphi_{[0,s]})| \, dA_s^n. \tag{2.33}$$

Then, by condition (iii), (3.31) holds.

Finally, we apply Kolmogorov's martingale inequality to obtain, taking into account (2.11) and (2.14), that

$$
\begin{aligned}
Q^{n,\lambda^0}&\left(\sup_{t\le 1}|M_t^n(\lambda^0)|>\delta\right)\\
&= Q^{n,\lambda^0}\left(\sup_{t\le 1}\left|n^{-1}\sum_{k=1}^{[nt]}\sigma_k(X^{n,k-1})[\xi_k-\mathsf{E}^{n,\lambda^0}(\xi_k|F_{k-1})]\right|>\delta\right)\\
&= Q^{n,\lambda^0}\left(\sup_{t\le 1}\left|n^{-1}\sum_{k=1}^{[nt]}\sigma_k(X^{n,k-1})[\xi_k-G'_\mu(\lambda_k^0\sigma_k(X^{n,k-1}))]\right|>\delta\right)\\
&\le \delta^{-2}\mathsf{E}^{n,\lambda^0}\left(n^{-1}\sum_{k=1}^{n}\sigma_k(X^{n,k-1})[\xi_k-G'_\mu(\lambda_k^0\sigma_k(X^{n,k-1}))]\right)^2\\
&\le \text{const}\,\delta^{-2}n^{-2}\sum_{k=1}^{n}\mathsf{E}^{n,\lambda^0}\sigma_k^2(X^{n,k-1})\,\mathsf{E}^{n,\lambda^0}(\xi_k^2|F_{k-1})\\
&\le \text{const}\,\delta^{-2}n^{-2}\sum_{k=1}^{n}\mathsf{E}^{n,\lambda^0}(\xi_k^2|F_{k-1}).
\end{aligned}
$$

It remains to observe that conditional expectation ξ^2 w.r.t $Q^{n,\lambda}$ can be expressed w.r.t P as

$$
\mathsf{E}^{n,\lambda^0}(\xi_k^2|F_{k-1}) = \mathsf{E}\,(\alpha_k(\lambda^0)\xi_k^2|F_{k-1}) = \frac{\mathsf{E}\,(\xi_k^2\exp(\lambda_k^0\sigma_k(X^{n,k-1})\xi_k)|F_{k-1})}{\mathsf{E}\,(\exp(\lambda_k^0\sigma_k(X^{n,k-1})\xi_k)|F_{k-1})}
$$

and is finite under Cramér's condition.

It completes the proof of the lemma.

Lower bound

To obtain lower bound we make use of the decomposition (2.11) to write $z^n(\lambda^0)$ (see (2.10)) in the following form:

$$
\begin{aligned}
z^n(\lambda^0) \;=\; \exp\Bigg(n\Bigg(&\int_0^1\lambda^0(s)\,\mathrm{d}M_s^n(\lambda^0)+\int_0^1\lambda^0(s)\,\mathrm{d}B_s^n(\lambda^0)\\
&-\int_0^1 G_\mu(\lambda^0(s)\,\sigma^n(X_{[0,s)}^n))\,\mathrm{d}A_s^n\Bigg)\Bigg),
\end{aligned}
\tag{2.34}
$$

with

$$
\int_0^1\lambda(s)\,\mathrm{d}M_s^n(\lambda) = n^{-1}\sum_{k=1}^{n}\lambda_k\sigma_k(X^{n,k-1})[\xi_k-\mathsf{E}^{n,\lambda}(\xi_k|F_{k-1})] \tag{2.35}
$$

and

$$
\int_0^1\lambda(s)\,\mathrm{d}B_s^n(\lambda) = n^{-1}\sum_{k=1}^{n}\lambda_k\sigma_k(X^{n,k-1})\,\mathsf{E}^{n,\lambda}(\xi_k|F_{k-1}).
$$

Next observe that in virtue of (2.14)

$$
\int_0^1\lambda^0(s)\,\mathrm{d}B_s^n(\lambda^0) = \int_0^1\lambda^0(s)\,\sigma^n(X_{[0,s)}^n)\,G'_\mu(\lambda^0(s)\sigma^n(X_{[0,s)}^n))\,\mathrm{d}A_s^n. \tag{2.36}
$$

Combining (2.34) and (2.36), we obtain that

$$
\begin{aligned}
z^n(\lambda^0) &= \exp\Bigg(n\Bigg(\int_0^1 \lambda^0(s)\, \mathrm{d}M_s^n(\lambda^0) + \int_0^1 \lambda^0(s)\,\sigma^n(X_{[0,s)}^n)\,G'_\mu(\lambda^0(s)\,\sigma^n(X_{[0,s)}^n))\,\mathrm{d}A_s^n \\
&\quad - \int_0^1 G_\mu(\lambda^0(s)\,\sigma^n(X_{[0,s)}^n))\,\mathrm{d}A_s^n\Bigg)\Bigg) \\
&= \exp\Bigg(n\Bigg(\int_0^1 \lambda^0(s)\, \mathrm{d}M_s^n(\lambda^0) + \int_0^1 \lambda^0(s)\,\sigma(\varphi_{[0,s]})\,G'_\mu(\lambda^0(s)\,\sigma(\varphi_{[0,s]}))\,\mathrm{d}A_s^n \\
&\quad - \int_0^1 G_\mu(\lambda^0(s)\,\sigma(\varphi_{[0,s]}))\,\mathrm{d}A_s^n + i_1^n + i_2^n\Bigg)\Bigg),
\end{aligned}
$$

with

$$
\begin{aligned}
i_1^n &= \int_0^1 \lambda^0(s)\,\sigma^n(X_{[0,s)}^n)\,G'_\mu(\lambda^0(s)\,\sigma^n(X_{[0,s)}^n))\,\mathrm{d}A_s^n \\
&\quad - \int_0^1 \lambda^0(s)\,\sigma(\varphi_{[0,s]})\,G'_\mu(\lambda^0(s)\,\sigma(\varphi_{[0,s]}))\,\mathrm{d}A_s^n, \\
i_2^n &= \int_0^1 G_\mu(\lambda^0(s)\,\sigma(\varphi_{[0,s]}))\,\mathrm{d}A_s^n - \int_0^1 G_\mu(\lambda^0(s)\,\sigma^n(X_{[0,s)}^n))\,\mathrm{d}A_s^n.
\end{aligned}
$$

Finally, by the choice of $\lambda^0(s)$ (see (2.19) and (2.24)) we write

$$
\begin{aligned}
z^n(\lambda^0) &= \exp\Bigg(n\Bigg(\int_0^1 \lambda^0(s)\, \mathrm{d}M_s^n(\lambda^0) \\
&\quad + \int_0^1 (\lambda^0(s)[\dot\varphi(s) - b(\varphi_{[0,s]})] - G_\mu(\lambda^0(s)\,\sigma(\varphi_{[0,s]})))\,\mathrm{d}A_s^n + i_1^n + i_2^n\Bigg)\Bigg) \\
&= \exp\Bigg(n\Bigg(\int_0^1 L(\sigma^{-1}(\varphi_{[0,s]})[\dot\varphi(s) - b(\varphi_{[0,s]})])\,\mathrm{d}A_s^n \\
&\quad + \int_0^1 \lambda^0(s)\,\mathrm{d}M_s^n(\lambda^0) + i_1^n + i_2^n\Bigg)\Bigg).
\end{aligned}
\tag{2.37}
$$

With this preparation we are ready to prove the following result.

Theorem 2.2. Let $\varphi = (\varphi(t))$, $0 \le t \le 1$, be absolutely continuous on $[0,1]$ with $\varphi(0) = x(= X_0^n)$ and

$$
J(\varphi) = \int_0^1 L(\sigma^{-1}(\varphi_{[0,s]})[\dot\varphi(s) - b(\varphi_{[0,s]})])\,\mathrm{d}s < \infty. \tag{2.38}
$$

Then under conditions of Theorem 2.1

$$
\liminf_\delta \liminf_n n^{-1} \ln \mathsf{P}\left(\sup_{t\le 1} |X_t^n - \varphi(t)| \le \delta\right) \ge -J(\varphi).
$$

Proof. We prove theorem in two step. Our first step is to show that (2.38) holds for continuously differentiable function φ.

Given a continuously differentiable function $\varphi(t)$ and λ^0, defined by (2.19), consider Q^{n,λ^0} and recall (see (2.7)) that

$$\mathrm{dP} = (z^n(\lambda^0))^{-1}\,\mathrm{d}Q^{n,\lambda^0}.$$

Thus, by (2.37) (I below is an indicator),

$$\begin{aligned}
\mathrm{P}\Big(\sup_{t\le 1}|X_t^n - \varphi(t)| \;\le\; \delta\Big)
&= \int_\Omega \mathrm{I}\Big(\sup_{t\le 1}|X_t^n - \varphi(t)| \le \delta\Big)(z^n(\lambda^0))^{-1}\,\mathrm{d}Q^{n,\lambda^0}\\
&= \mathrm{E}^{n,\lambda^0}\,\mathrm{I}\Big(\sup_{t\le 1}|X_t^n - \varphi(t)| \le \delta\Big)\\
&\quad\times \exp\Big(-n\Big(\int_0^1 L(\sigma^{-1}(\varphi_{[0,s]})[\dot\varphi(s) - b(\varphi_{[0,s]})])\,\mathrm{d}A_s^n\\
&\quad+\int_0^1 \lambda^0(s)\,\mathrm{d}M_s^n(\lambda^0) + i_1^n + i_2^n\Big)\Big)\\
&\ge \mathrm{E}^{n,\lambda^0}\,\mathrm{I}\Big(\sup_{t\le 1}|X_t^n - \varphi(t)| \le \delta\Big)\\
&\quad\times \exp\Big(-n\Big(\int_0^1 L(\sigma^{-1}(\varphi_{[0,s]})[\dot\varphi(s) - b(\varphi_{[0,s]})])\,\mathrm{d}s\Big)\Big)\\
&\quad\times \exp\Big(-n\Big(\int_0^1 L(\sigma^{-1}(\varphi_{[0,s]})[\dot\varphi(s) - b(\varphi_{[0,s]})])\,\mathrm{d}(A_s^n - s)\\
&\quad+\Big|\int_0^1 \lambda^0(s)\,\mathrm{d}M_s^n(\lambda^0)\Big| + |i_1^n| + |i_2^n|\Big)\Big).
\end{aligned}\tag{2.39}$$

Hence, we have to estimate the second exponent. To this end, we first evaluate i_1^n and i_2^n. If one repeats slightly modified calculations (2.25)–(2.28) and (2.32)–(2.33) of Lemma 2.1 one can easily obtain by the similar argument the following estimations:

$$\begin{aligned}
|i_1^n| &= \Big|\int_0^1 \lambda^0(s)\,\sigma^n(X^n_{[0,s)})\,G'_\mu(\lambda^0(s)\,\sigma^n(X^n_{[0,s)}))\,\mathrm{d}A_s^n\\
&\quad-\int_0^1 \lambda^0(s)\,\sigma(\varphi_{[0,s]})\,G'_\mu(\lambda^0(s)\,\sigma(\varphi_{[0,s]}))\,\mathrm{d}A_s^n\Big|\\
&\le \text{const}\,\sup_{t\le 1}|X_t^n - \varphi(t)| + \gamma_n
\end{aligned}$$

and

$$|i_2^n| = \Big|\int_0^1 G_\mu(\lambda^0(s)\,\sigma(\varphi_{[0,s]}))\,\mathrm{d}A_s^n - \int_0^1 G_\mu(\lambda^0(s)\,\sigma^n(X^n_{[0,s)}))\,\mathrm{d}A_s^n\Big|$$

$$\leq \text{ const } \sup_{t\leq 1} |X_t^n - \varphi(t)| + \gamma_n$$

with

$$\gamma_n = \int_0^1 |\sigma^n(\varphi_{[0,s]}) - \sigma(\varphi_{[0,s]})| \, dA_s^n.$$

Hence, on the set $\left(\sup_{t\leq 1} |X_t^n - \varphi(t)| \leq \delta\right)$ we have

$$|i_1^n| \leq \text{const } \times \delta + \gamma_n \tag{2.40}$$

and

$$|i_2^n| \leq \text{const } \times \delta + \gamma_n. \tag{2.41}$$

We next define the set

$$C_{\delta,q} = \left(\sup_{t\leq 1} |X_t^n - \varphi(t)| \leq \delta\right) \bigcap \left(\left|\int_0^1 \lambda^0(s) \, dM_s^n(\lambda^0)\right| \leq \delta\right) \tag{2.42}$$

Then, by combining (2.39) and (2.40)–(2.42) we obtain that

$$\begin{aligned}
\mathsf{P}&\left(\sup_{t\leq 1} |X_t^n - \varphi(t)| \leq \delta\right) \\
&\geq \exp\Bigg(-n \int_0^1 L(\sigma^{-1}(\varphi_{[0,s]})[\dot\varphi(s) - b(\varphi_{[0,s]})]) \, ds \\
&\qquad -n \int_0^1 L(\sigma^{-1}(\varphi_{[0,s]})[\dot\varphi(s) - b(\varphi_{[0,s]})]) \, d(A_s^n - s)\Bigg) \\
&\qquad \times \int_{C_{\delta,q}} \exp\left(-n\left(\left|\int_0^1 \lambda^0(s) \, dM_s^n(\lambda^0)\right| + |i_1^n| + |i_2^n|\right)\right) dQ^{n,\lambda^0} \\
&\geq \exp\Bigg(-n \int_0^1 L(\sigma^{-1}(\varphi_{[0,s]})[\dot\varphi(s) - b(\varphi_{[0,s]})]) \, ds \\
&\qquad -n \int_0^1 L(\sigma^{-1}(\varphi_{[0,s]})[\dot\varphi(s) - b(\varphi_{[0,s]})]) \, d(A_s^n - s)\Bigg) \\
&\qquad \times \int_{C_{\delta,q}} \exp\left(-n(q + \text{const } \times \delta + 2\gamma_n)\right) dQ^{n,\lambda^0}
\end{aligned}$$

and so

$$\begin{aligned}
n^{-1} \ln \mathsf{P}&\left(\sup_{t\leq 1} |X_t^n - \varphi(t)| \leq \delta\right) \\
&\geq -\int_0^1 L(\sigma^{-1}(\varphi_{[0,s]})[\dot\varphi(s) - b(\varphi_{[0,s]})]) \, ds \\
&\quad -\int_0^1 L(\sigma^{-1}(\varphi_{[0,s]})[\dot\varphi(s) - b(\varphi_{[0,s]})]) \, d(A_s^n - s) \\
&\quad -q - \text{const } \times \delta - 2\gamma_n + n^{-1} \ln Q^{n,\lambda^0}(C_{\delta,q}).
\end{aligned}$$

Since, by Lemma 1.6 of Section 1.5

$$\int_0^1 L(\sigma^{-1}(\varphi_{[0,s]})[\dot\varphi(s) - b(\varphi_{[0,s]})])\ \mathrm{d}(A_s^n - s)$$

tends to 0 as n $\to \infty$; by condition (iii), γ_n tends to 0 as n $\to \infty$, and q, δ are arbitrary, it remains to prove that

$$\lim_n Q^{n,\lambda^0}(C_{\delta,q}) = 1$$

for every $\delta > 0$ and $q > 0$.

To this end, note that

$$Q^{n,\lambda^0}(C_{\delta,q}) \geq 1 - Q^{n,\lambda^0}\left(\sup_{t\leq 1}|X_t^n - \varphi(t)| > \delta\right) - Q^{n,\lambda^0}\left(\left|\int_0^1 \lambda^0(s)\ \mathrm{d}M_s^n(\lambda^0)\right| > \delta\right),$$

and so we have to prove that for every $\delta > 0$ and $q > 0$

$$\lim_n Q^{n,\lambda^0}\left(\sup_{t\leq 1}|X_t^n - \varphi(t)| > \delta\right) = 0, \tag{2.43}$$

$$\lim_n Q^{n,\lambda^0}\left(\left|\int_0^1 \lambda^0(s)\ \mathrm{d}M_s^n(\lambda^0)\right| > \delta\right) = 0. \tag{2.44}$$

Evidently, (2.43) is the statement of Lemma 2.1 and so we need only prove (2.44). To this end, we apply Chebyshov's inequality and obtain, taking into account (2.14) and (2.35), the following estimation:

$$\begin{aligned} Q^{n,\lambda^0}&\left(\left|\int_0^1 \lambda^0(s)\ \mathrm{d}M_s^n(\lambda^0)\right| > \delta\right) \\ &\leq \delta^{-2}\mathsf{E}^{n,\lambda^0}\left(n^{-1}\sum_{k=1}^n \lambda_k^0\sigma_k(X^{n,k-1})[\xi_k - G'_\mu(\lambda_k^0\sigma_k(X^{n,k-1}))]\right)^2 \\ &\leq \text{const}\,\delta^{-2}n^{-2}\sup_{s\leq 1}\lambda^0(s)\sum_{k=1}^n \mathsf{E}^{n,\lambda^0}\sigma_k^2(X^{n,k-1})\,\mathsf{E}^{n,\lambda^0}(\xi_k^2|F_{k-1}). \end{aligned}$$

Then, the argument used for proving Lemma 2.1 in this section leads to the conclusion that (2.44) holds.

This completes the first step. It remains to prove the statement of theorem for absolutely continuous in [0,1] functions with $J(\varphi) < \infty$. Given absolutely continuous $\varphi(t)$, $\varphi(0) = x$, there exists a sequence of continuously differentiable functions $\varphi^N = (\varphi^N(t))$, $N \geq 1$, such that

$$\lim_n \sup_{t\leq 1}\left|\varphi(t) - \varphi^N(t)\right| = 0. \tag{2.45}$$

By applying the result of the first step we conclude that

$$\liminf_\delta\ \liminf_n n^{-1}\ln\mathsf{P}\left(\sup_{t\leq 1}\left|X_t^n - \varphi^N(t)\right| \leq \delta\right) \geq -J(\varphi^N).$$

We next note that by the triangle inequality

$$\mathsf{P}\left(\sup_{t\leq 1}|X_t^n-\varphi(t)|\leq\delta+\sup_{t\leq 1}\left|\varphi(t)-\varphi^N(t)\right|\right)\geq\mathsf{P}\left(\sup_{t\leq 1}\left|X_t^n-\varphi^N(t)\right|\leq\delta\right). \tag{2.46}$$

Then, bringing (2.45) and (2.46) together, we obtain

$$\liminf_{\delta}\ \liminf_{n} n^{-1}\ln\mathsf{P}\left(\sup_{t\leq 1}|X_t^n-\varphi(t)|\leq\delta\right)\geq-\liminf_{N} J(\varphi^N).$$

It is then an easy step to show that

$$\liminf_{N} J(\varphi^N)=J(\varphi).$$

This completes the proof of Theorem 2.2.

Upper bound

Theorem 2.3. Let Cramér's condition and conditions (i)-(iii) of Theorem 2.1 hold. Then, for all $\varphi\in D[0,1]$ with $\varphi_0=x$,

$$\limsup_{\delta\to 0}\ \limsup_{n} n^{-1}\ln\mathsf{P}\left(\sup_{t\leq 1}|X_t^n-\varphi(t)|\leq\delta\right)\leq -I(\varphi)$$

with I defined in Theorem 2.1.

Proof. Consider a step-function $(\lambda(t))_{t\leq 1}$ of the form

$$\lambda(t)=\sum_i\lambda_{t_i}\,\mathrm{I}_{]t_{i-1},t_i]},$$

and define

$$\int_0^1\lambda(t)\,\mathrm{d}\varphi(t)=\sum_i\lambda_{t_i}[\varphi_{1\wedge t_i}-\varphi_{1\wedge t_{i-1}}].$$

Recall that for any function $\lambda(t)$ and $\Delta=1/n$

$$\int_0^1\lambda(s)\,\mathrm{d}X_s^n=\sum_{k=1}^n\lambda(k\Delta)(X_k^n-X_{k-1}^n).$$

Consider $z^n(\lambda)$, defined by (2.10) with step-function $\lambda(t)$, and recall that $\mathsf{E}z^n(\lambda)=1$. Then, for $\varphi\in D[0,1]$

$$\begin{aligned}1\ \geq\ &\mathsf{E}\,\mathrm{I}\left(\sup_{t\leq 1}|X_t^n-\varphi(t)|\leq\delta\right)\ \exp\Bigg(n\Bigg[\int_0^1\lambda(s)\,\mathrm{d}X_s^n-\int_0^1\lambda(s)\,b^n(X_{[0,s)}^n)\,\mathrm{d}A_s^n\\&-\int_0^1 G_\mu(\lambda(s)\,\sigma^n(X_{[0,s)}^n))\,\mathrm{d}A_s^n\Bigg]\Bigg)\\\geq\ &\mathsf{E}\,\mathrm{I}\left(\sup_{t\leq 1}|X_t^n-\varphi(t)|\leq\delta\right)\ \exp\Bigg(n\Bigg[\int_0^1\lambda(s)\,\mathrm{d}\varphi(s)-\int_0^1\lambda(s)\,b(\varphi_{[0,s]})\,\mathrm{d}s\\&-\int_0^1 G_\mu(\lambda(s)\,\sigma(\varphi_{[0,s]}))\,\mathrm{d}s-|i_1|-|i_2|-|i_3|-|i_4|-|i_5|\Bigg]\Bigg),\end{aligned}$$

with

$$i_1 = \int_0^1 \lambda(s)\, \mathrm{d}(X_s^n - \varphi(s)),$$

$$i_2 = \int_0^1 \lambda(s)[b(\varphi_{[0,s]}) - b^n(X_{[0,s)}^n)]\, \mathrm{d}A_s^n,$$

$$i_3 = \int_0^1 [G_\mu(\lambda(s)\,\sigma(\varphi_{[0,s]})) - G_\mu(\lambda(s)\,\sigma^n(X_{[0,s)}^n))]\, \mathrm{d}A_s^n,$$

$$i_4 = \int_0^1 \lambda(s)\, b(\varphi_{[0,s]})\, \mathrm{d}\,(s - A_s^n),$$

$$i_5 = \int_0^1 G_\mu(\lambda(s)\,\sigma(\varphi_{[0,s]}))\, \mathrm{d}\,(s - A_s^n).$$

Define

$$\tilde{I}(\varphi) = \sup \left[\int_0^1 \lambda(s)\, \mathrm{d}\varphi(s) - \int_0^1 \lambda(s)\, b(\varphi_{[0,s]})\, \mathrm{d}s - \int_0^1 G_\mu(\lambda(s)\,\sigma(\varphi_{[0,s]}))\, \mathrm{d}s \right],$$

where sup is taken over all step-functions λ. Then from the last inequality we have that

$$1 \geq \exp\,(n[\tilde{I}(\varphi) - |i_1| - |i_2| - |i_3| - |i_4| - |i_5|])\, \mathsf{P}\left(\sup_{t \leq 1} |X_t^n - \varphi(t)| \leq \delta \right).$$

By repeating the argument, used for proving the lower bound, it is an easy step to show that for any $\varepsilon > 0$ one can choose $\delta_\varepsilon > 0$ such that on the set $(\sup_{t \leq 1} |X_t^n - \varphi(t)| \leq \delta)$ with $\delta \leq \delta_\varepsilon$

$$|i_1| + |i_2| + |i_3| + |i_4| + |i_5| \leq \varepsilon,$$

and so

$$n^{-1} \ln \mathsf{P}\left(\sup_{t \leq 1} |X_t^n - \varphi(t)| \leq \delta \right) \leq -\tilde{I}(\varphi) + \varepsilon.$$

Since ε is arbitrary we arrive at the following upper bound:

$$\limsup_{\delta \to 0}\ \limsup_{n}\, n^{-1} \ln \mathsf{P}\left(\sup_{t \leq 1} |X_t^n - \varphi(t)| \leq \delta \right) \leq -\tilde{I}(\varphi).$$

It remains to show that $\tilde{I} = I$. If φ is not absolutely continuous on $[0, 1]$, it is an easy step to show that $\tilde{I}(\varphi) = \infty$. Let φ be an absolutely continuous function. By the definition, $\tilde{I}(\varphi) \leq I(\varphi)$. Thus, one has to prove that $\tilde{I}(\varphi) \geq I(\varphi)$. The reader is referred to Liptser and Pukhalskii (1992) (Lemma 6.1) for the proof of this statement.

Exponential tightness

According to the scheme of Section 1.4, for proving main result it remains to check the exponential tightness of the sequence X^n. We first verify that

$$\lim_{c\to\infty} \limsup_n n^{-1} \ln \mathsf{P}\left(\sup_{t\le 1} |X_t^n| \ge c\right) = -\infty.$$

To this end, define the stopping time

$$\theta_c = \inf\,(t:\ |X_t^n| \ge c) \wedge 1,$$

and note that

$$\left(\omega:\ \sup_{t\le 1} |X_t^n| \ge c\right) = (\omega:\ |X_{\theta_c}^n| \ge c) = (X_{\theta_c}^n \ge c) \cup (X_{\theta_c}^n \le -c). \tag{2.47}$$

Hence,

$$\mathsf{P}\left(\sup_{t\le 1} |X_t^n| \ge c\right) \le 2\max\,[\mathsf{P}(X_{\theta_c}^n \ge c), \mathsf{P}(X_{\theta_c}^n \le -c)]. \tag{2.48}$$

Therefore, it is sufficient to prove that

$$\lim_{c\to\infty} \limsup_n n^{-1} \ln \mathsf{P}\,(X_{\theta_c}^n \ge c) = \lim_{c\to\infty} \limsup_n n^{-1} \ln \mathsf{P}\,(X_{\theta_c}^n \le -c) = -\infty. \tag{2.49}$$

Recall that for any $\lambda(s)$ the process $z_t^n(\lambda)$ (see (2.4) and (2.10)) is a martingale w.r.t. filtration $(F_t)_{t\le 1}$ with $F_t \equiv F_{[nt]}$, and $\mathsf{E}z^n(\lambda) = 1$. Pick $\lambda(s) \equiv r \ge 0$. Then, for the stopping time θ_c, by martingale property, we may write (see (2.10)):

$$1 \le \mathsf{E}\,I(X_{\theta_c}^n \ge c)\exp\left(n\left[\int_0^{\theta_c} r\,\mathrm{d}X_s^n - \int_0^{\theta_c} rb^n(X_{[0,s)}^n)\,\mathrm{d}A_s^n - \int_0^{\theta_c} G_\mu(r\sigma^n(X_{[0,s)}^n))\,\mathrm{d}A_s^n\right]\right). \tag{2.50}$$

Note that, since $\mathsf{E}\xi_1 = 0$, by Jensen's inequality, $G_\mu(\beta) = \ln \mathsf{E}e^{\beta\xi_1} \ge 0$. Assume, for the simplicity, that $X_0^n = 0$. Then, by boundedness condition (i) of Theorem 2.1, on the set $(X_{\theta_c}^n \ge c)$

$$\begin{aligned}
&\int_0^{\theta_c} r\,\mathrm{d}X_s^n - \int_0^{\theta_c} rb^n(X_{[0,s)}^n)\,\mathrm{d}A_s^n - \int_0^{\theta_c} G_\mu(r\sigma^n(X_{[0,s)}^n))\,\mathrm{d}A_s^n \\
&\quad\ge rc - \int_0^{\theta_c} r|b^n(X_{[0,s)}^n)|\,\mathrm{d}A_s^n - \int_0^{\theta_c} G_\mu(r\sigma^n(X_{[0,s)}^n))\,\mathrm{d}A_s^n \\
&\quad\ge rc - rK - G_\mu^*(rk),
\end{aligned} \tag{2.51}$$

where

$$G_\mu^*(rK) = G_\mu(rK) \vee G_\mu(-rK).$$

Tying (2.50) and (2.51) together we arrive at the following inequality

$$1 \geq \mathsf{E}\,I(X^n_{\theta_c} \geq c)\ \exp\left(n[rc - rK - G^*_\mu(rK)]\right),$$

and so

$$n^{-1} \ln \mathsf{P}(X^n_{\theta_c} \geq c) \leq -\sup_{r \geq 0} \left[r(c-K) - G^*_\mu(rK)\right].$$

Suppose, for the definiteness, that $G_\mu(rK) \geq G_\mu(-rK)$. Then, by property (iii) of Lemma 1.1 in Section 1.1,

$$n^{-1} \ln \mathsf{P}(X^n_{\theta_c} \geq c) \leq -L\left(\frac{c-K}{K}\right),$$

where $L(y) = \sup_{\beta \in R}[\beta y - G_\mu(\beta)]$. Since, by property (iv) of Lemma 1.1 in Section 1.1 $L(K^{-1}(c-K)) \to \infty$ as $c \to \infty$ we get

$$\lim_{c \to \infty} \limsup_n n^{-1} \ln \mathsf{P}(X^n_{\theta_c} \geq c) \leq -\infty.$$

The first relation in (2.49) is proved. The second relation in (2.49) is proved similarly.

We next verify that

$$\lim_{\delta \to 0} \limsup_n \sup_{\tau \leq 1} n^{-1} \ln \mathsf{P}\left(\sup_{t \leq \delta} |X^n_{t+\tau} - X^n_\tau| \geq \eta\right) = -\infty. \tag{2.52}$$

For any stopping τ ($\tau \leq 1$) we define a process

$$x^n_t = X^n_{t+\tau} - X^n_\tau,$$

and put

$$\gamma_\eta = \inf\,(t\colon\ |x^n_t| \geq \eta) \wedge \delta.$$

Evidently, γ_η is a stopping time w.r.t. the filtration $(F_{t+\tau})_{t \leq 1}$ with $F_{\tau+t} \equiv F_{[n(\tau+t)]}$. Then

$$\begin{aligned}\left(\omega\colon \sup_{t \leq \delta} |X^n_{t+\tau} - X^n_\tau| \geq \eta\right) &= \left(\sup_{t \leq \delta} |x^n_t| \geq \eta\right)\\ &= \left(\sup_{t \leq \delta} x^n_t \geq \eta\right) \bigcup \left(\sup_{t \leq \delta} x^n_t \leq -\eta\right),\end{aligned}$$

and so

$$\mathsf{P}\left(\sup_{t \leq \delta} |X^n_{t+\tau} - X^n_\tau| \geq \eta\right) \leq 2\max\left(\mathsf{P}(x^n_{\gamma_\eta} \geq \eta), \mathsf{P}(x^n_{\gamma_\eta} \leq -\eta)\right).$$

Hence, (2.52) will be proved, once we show that

$$\begin{aligned}&\lim_{\delta \to 0} \limsup_n n^{-1} \ln \sup_{\tau \leq 1} \mathsf{P}(x^n_\gamma \geq \eta)\\ &\qquad = \lim_{\delta \to 0} \limsup_n n^{-1} \ln \sup_{\tau \leq 1} \mathsf{P}(x^n_\gamma \leq -\eta) = -\infty.\end{aligned} \tag{2.53}$$

We next note that $z^n_{t+\tau}(\lambda)$ is a martingale w.r.t. the filtration $(F_{t+\tau})$, $\tau \geq 0$ and so

$$\frac{\mathsf{E}\,(z^n_{\tau+\gamma}(\lambda)|F_\tau)}{z^n_\tau(\lambda)} = 1.$$

The last equality implies (with $\lambda(s) \equiv r \geq 0$) that

$$\begin{aligned}
1 \;\geq\; & \mathsf{E}\,\mathrm{I}(x^n_\gamma \geq \eta)\, \exp\Bigg(n\Bigg[\int_\tau^{\tau+\gamma} r \,\mathrm{d}X^n_s - \int_\tau^{\tau+\gamma} r b^n(X^n_{[0,s)})\,\mathrm{d}A^n_s \\
& - \int_\tau^{\tau+\gamma} G_\mu(r\sigma^n(X^n_{[0,s)}))\,\mathrm{d}A^n_s\Bigg]\Bigg) \\
\geq\; & \mathsf{E}\,\mathrm{I}(x^n_\gamma \geq \eta)\, \exp\Bigg(n\Bigg[r(X^n_{\tau+\gamma} - X^n_\tau) \\
& - \int_\tau^{\tau+\delta} r|b^n(X^n_{[0,s)})|\,\mathrm{d}A^n_s - \int_\tau^{\tau+\delta} G^*_\mu(rK)\,\mathrm{d}A^n_s\Bigg]\Bigg) \\
\geq\; & \mathsf{E}\,\mathrm{I}(x^n_\gamma \geq \eta)\, \exp\left(n\left[r\eta - rK\frac{[n\delta]}{n} - G^*_\mu(rK)\frac{[n\delta]}{n}\right]\right).
\end{aligned}$$

Denote $\delta_n = [n\delta]/n$ and observe that $\delta_n \to \delta$ as $n \to \infty$. Then,

$$n^{-1}\ln \sup_{\tau\leq 1} \mathsf{P}\,(x^n_\gamma \geq \eta) \leq \sup_{r\geq 0}\left[r(\eta - K\delta_n) - \delta_n G^*_\mu(rK)\right].$$

Assuming again that $G_\mu(rK) \geq G_\mu(-rK)$ we get

$$n^{-1}\ln \sup_{\tau\leq 1} \mathsf{P}\,(x^n_\gamma \geq \eta) \leq -\delta_n L\left(\frac{\eta - \delta_n K}{K\delta_n}\right).$$

Hence, by property (iv) of Lemma 1.1 in Section 1.1,

$$\lim_{\delta\to 0}\,\limsup_n n^{-1}\ln \sup_{\tau\leq 1} \mathsf{P}\,(x^n_\gamma \geq \eta) = -\infty.$$

The second relation in (2.53) is proved similarly.

Comments

The problem of large deviations for continuous time Markov processes of the form (2.3) was considered by Wentzell and Freidlin (see Wentzell and Freidlin (1984)). The case of processes with independent increments was considered in pioneer work by Borovkov (1967) and Mogulskii (1976). For the diffusion-type processes the results have been generalized by a number of authors (see Azencott (1980), Baldi and Chaleyat-Maurel (1988) and Narita (1988)). The case of non-homogeneous diffusion with the coefficients depending on the past was considered by Gutland (1988) and Liptser and Pukhalskii (1992). We restrict ourself to the case of bounded coefficients; however, this result may be proved under conditions of linear grouth. The result of this chapter for the discrete time processes obtained by Gulinsky seems to be new.

Chapter 3

Large deviations for the recursive scheme with stationary disturbances

3.1. LARGE DEVIATIONS FOR THE SUMS OF STATIONARY SEQUENCE WITH THE WOLD-TYPE REPRESENTATION

Setting. Main idea
Let (ξ_k), $-N < k < \infty$, $\xi_k \in R$, be a sequence of i.i.d. random variables with a common distribution μ, $\mathsf{E}\xi_1 = 0$, $\mathsf{E}\xi_1^2 = 1$. We assume that (ξ_k) satisfies Cramér's condition, i.e.,

$$\mathsf{E}\exp(\lambda\xi_1) < \infty,$$

for all $\lambda \in R$ and the logarithmic moment generating function

$$G_\mu(\lambda) = \ln \mathsf{E}\exp(\lambda\xi_1)$$

satisfies the regularity condition

$$\lim_{|\lambda|\to\infty} \frac{G_\mu(\lambda)}{|\lambda|} = \infty.$$

Let (Y_k), $-N < k < \infty$, be a sequence of random variables such that Y_k admits the Wold-type decomposition

$$Y_k = \sum_{j=-N}^{k} a(k-j)\xi_k \tag{3.1}$$

with a numerical sequence $(a(i))$, $i \geq 0$.

Note that if $N = +\infty$ and $\sum_{k\geq 0} a^2(k) < \infty$, then (Y_k), $-\infty < k < \infty$, is a strictly stationary sequence. Our first goal is to derive the large deviation principle for the sums (S_n/n), $n \geq 1$, with

$$S_n = \sum_{k=1}^{n} Y_k. \tag{3.2}$$

Define also the family of processes $W^n = (W_t^n)$, $0 \leq t \leq 1$, with

$$W_t^n = n^{-1}\sum_{k=1}^{[nt]} Y_k, \tag{3.3}$$

where $Y = (Y_k)$, $-\infty < k < \infty$, is a strictly stationary sequence with

$$Y_k = \sum_{j=-\infty}^{k} a(k-j)\xi_k.$$

In this case we will be interested in an asymptotic behaviour of the probability

$$\mathsf{P}\left(\sup_{t\leq 1}|W_t^n-\varphi(t)|\leq\delta\right),$$

for any $\varphi\in C[0,1]$.

To study the problem we make use an approach based on the following decomposition (see Liptser and Shiryaev (1989)):

$$S_n = M_n + V_1 - V_{n+1},$$

with a square integrable martingale $M = (M_n)$, $n\geq 1$, and a strictly stationary sequence (V_n), $n\geq 0$. The decomposition takes place under some weak dependence condition for (Y_k). We will show that this condition in the terms of the sequence $a(i)$ can be expressed as

$$\sum_{i=1}^{\infty} i|a(i)| < \infty. \tag{3.4}$$

Throughout this chapter we make use the following idea. Let

$$S_n = M_n + U_n, \qquad n\geq 1, \tag{3.5}$$

and suppose that (M_n/n), $n\geq 1$, satisfies the local large deviations principle with rate function I, i.e.,

$$\limsup_{\delta\to 0}\ \limsup_{n\to\infty} n^{-1}\ln\mathsf{P}\left(|M_n/n-y|\leq\delta\right)\geq -I(y), \tag{3.6}$$

$$\liminf_{\delta\to 0}\ \liminf_{n\to\infty} n^{-1}\ln\mathsf{P}\left(|M_n/n-y|\leq\delta\right)\leq -I(y). \tag{3.7}$$

Assume that for each $q>0$

$$\limsup_{n\to\infty} n^{-1}\ln\mathsf{P}\left(|U_n|>nq\right) = -\infty. \tag{3.8}$$

Then, the sequence (S_n/n), $n\geq 1$, satisfies the local large deviations principle with the same rate function.

We formulate and prove this result in a function space context.

Let $X^n = (X_t^n)$, $0\leq t\leq 1$, be a family of processes with trajectories in $D[0,1]$, which admit a representation

$$X_t^n = M_t^n + U_t^n, \tag{3.9}$$

and a sequence $M^n = (M_t^n)$, $n\geq 1$, satisfies the C-local large deviation principle with rate function I, i.e., for any $\varphi\in C[0,1]$

$$-\liminf_{\delta\to 0}\ \liminf_{n\to\infty} n^{-1}\ln\mathsf{P}\left(\sup_{0\leq t\leq 1}|M_t^n-\varphi(t)|\leq\delta\right)$$

$$= -\limsup_{\delta\to 0}\ \limsup_{n\to\infty} n^{-1}\ln\mathsf{P}\left(\sup_{0\leq t\leq 1}|M_t^n-\varphi(t)|\leq\delta\right) = I(\varphi). \tag{3.10}$$

Lemma 3.1. Assume that for every $q > 0$,

$$\limsup_{n\to\infty} n^{-1} \ln \mathsf{P}\left(\sup_{0\le t\le 1} |U_t^n| > q \right) = -\infty. \tag{3.11}$$

Then, X^n satisfies the local large deviations principle with the same rate function I.

Proof. Upper bound. In view of (3.9), by the triangle inequality,

$$\sup_{0\le t\le 1} |M_t^n - \varphi(t)| \le \sup_{0\le t\le 1} |X_t^n - \varphi(t)| + \sup_{0\le t\le 1} |U_t^n|.$$

Hence,

$$\begin{aligned}
&\mathsf{P}\left(\sup_{0\le t\le 1} |X_t^n - \varphi(t)| \le \delta \right) \\
&\quad \le \mathsf{P}\left(\sup_{0\le t\le 1} |M_t^n - \varphi(t)| \le \delta + \sup_{0\le t\le 1} |U_t^n| \right) \\
&\quad \le \mathsf{P}\left(\sup_{0\le t\le 1} |M_t^n - \varphi(t)| \le \delta + q \right) + \mathsf{P}\left(\sup_{0\le t\le 1} |U_t^n| > q \right) \\
&\quad \le 2\max\left[\mathsf{P}\left(\sup_{0\le t\le 1} |M_t^n - \varphi(t)| \le \delta + q \right), \mathsf{P}\left(\sup_{0\le t\le 1} |U_t^n| > q \right) \right],
\end{aligned}$$

and so (pick $q = \delta$)

$$\begin{aligned}
&\limsup_{\delta\to 0} \limsup_{n} n^{-1} \ln \mathsf{P}\left(\sup_{0\le t\le 1} |X_t^n - \varphi(t)| \le \delta \right) \\
&\qquad \le 2\left(-I(\varphi) \vee \limsup_{n} n^{-1} \ln \mathsf{P}\left(\sup_{0\le t\le 1} |U_t^n| > q \right) \right).
\end{aligned}$$

Thus, in view of (3.11) we obtain (3.10).

Lower bound. By repeating the triangle inequality argument, we get

$$\begin{aligned}
\mathsf{P}\left(\sup_{0\le t\le 1} |X_t^n - \varphi(t)| \le \delta \right) &\ge \mathsf{P}\left(\sup_{0\le t\le 1} |M_t^n - \varphi(t)| \le \delta - \sup_{0\le t\le 1} |U_t^n| \right) \\
&\ge \mathsf{P}\left(\sup_{0\le t\le 1} |M_t^n - \varphi(t)| \le \delta/2 \right) \\
&\quad - \mathsf{P}\left(\sup_{0\le t\le 1} |U_t^n| > \delta/2 \right).
\end{aligned}$$

Thus,

$$\begin{aligned}
&2\max\left[\mathsf{P}\left(\sup_{0\le t\le 1} |X_t^n - \varphi(t)| \le \delta \right), \mathsf{P}\left(\sup_{0\le t\le 1} |U_t^n| > \delta/2 \right) \right] \\
&\qquad \ge \mathsf{P}\left(\sup_{0\le t\le 1} |M_t^n - \varphi(t)| \le \delta/2 \right),
\end{aligned}$$

and so

$$\liminf_{\delta} \liminf_{n} n^{-1} \ln \mathsf{P}\left(\sup_{0 \le t \le 1} |X_t^n - \varphi(t)| \le \delta \right)$$
$$\vee \limsup_{\delta} \limsup_{n} n^{-1} \ln \mathsf{P}\left(\sup_{0 \le t \le 1} |U_t^n| > \delta/2 \right) \ge -I(\varphi).$$

Therefore, by (3.11) we arrive at the lower bound.

Remark 3.1. Under our condition the passage from the C-local to full large deviation principle is evident and so later on we will consider only local one.

Representation of strictly stationary processes
We now turn back to decomposition (3.5) and discuss a weak dependence assumption we need to impose on a stationary process. Later on in this paragraph we follow in our treatment Liptser and Shiryaev (1989).

Consider (Y_k), $-\infty < k < \infty$, and define a family of σ-algebras $F = (F_k)$, $k \ge 1$, with $F_k = \sigma(\ldots, Y_0, \ldots, Y_k)$. Denote

$$\gamma_k(p) = (\mathsf{E}\,|\mathsf{E}\,(Y_k | F_0)|^{p/p-1})^{(p-1)/p}. \tag{3.12}$$

Assume that for some $p \ge 2$, $\mathsf{E}\,|Y_0|^p < \infty$. In this case, it is easy to show with the help of Jensen's inequality that

$$\gamma_{k+1}(p) \le \gamma_k(p), \qquad k \ge 1.$$

Hence, we may consider $\gamma_k(p)$ as some weak dependence coefficient. In what follows an important role will play the following condition:

$$\sum_k \gamma_k(p) < \infty. \tag{3.13}$$

The weak dependence property of stationary processes is traditionally expressed in terms of mixing coefficients. In this regard we now recall some definitions and give conditions, expressed in terms of corresponding mixing coefficients, which guarantee (3.13) (see also Chapter 5 in this book).

Denote

$$F^k = \sigma(Y_k, Y_{k+1}, \ldots).$$

The parameter φ_k, $k \ge 1$, with

$$\varphi_k = \sup_{A \in F_0, B \in F^k} |\mathsf{P}(B/A) - \mathsf{P}(B)|,$$

$\mathsf{P}(A) > 0$, is called the uniformly strong mixing coefficient.

The parameter α_k, $k \ge 1$, with

$$\alpha_k = \sup_{A \in F_0, B \in F^k} |\mathsf{P}(AB) - \mathsf{P}(A)\,\mathsf{P}(B)|$$

is called the strong mixing coefficient.

We give without proof the following inequalities:

$$\gamma_k(p) \leq 2\varphi_k^{(p-1)/p}(\mathsf{E}\,|Y_0|^p)^{1/p}, \qquad p \geq 2,$$

$$\gamma_k(p) \leq 2\,(2^{1/p} + 1)\alpha_k^{(p-2)/p}\mathsf{E}\,(|Y_0|^p)^{1/p}, \qquad p \geq 2.$$

It is then an easy matter to check that (3.13) follows from the inequalities

$$\sum_{k\geq 1} \varphi_k^{(p-1)/p} < \infty, \qquad p \geq 2,$$

$$\sum_{k\geq 1} \alpha_k^{(p-2)/p} < \infty, \qquad p \geq 2.$$

Let for some $p > 2$, $\mathsf{E}\,|Y_0|^p < \infty$ and (3.13) hold. It turns out that in this case

$$\sum_{k\geq n} \mathsf{E}\,|\mathsf{E}\,(Y_k|F_{n-1})| < \infty,$$

$$\sum_{k\geq n} \mathsf{E}\,|\mathsf{E}\,(Y_k|F_n)| < \infty$$

and we can define random variables

$$V_n = \sum_{i\geq n} \mathsf{E}\,(Y_i|F_{n-1}),$$

$$\eta_n = \sum_{i\geq n} [\mathsf{E}\,(Y_i|F_n) - \mathsf{E}\,(Y_i|F_{n-1})].$$

One can check by direct calculation that

$$Y_n = \eta_n + V_n - V_{n+1} \tag{3.14}$$

and

$$\frac{S_n}{n} = n^{-1}\sum_{k=1}^{n} Y_k = n^{-1}\sum_{k=1}^{n} \eta_k + n^{-1}(V_1 - V_{n+1}).$$

Clearly, (V_k), $k \geq 1$, is a strictly stationary sequence and (η_k, F_k) is a sequence of martingale differences:

$$\begin{aligned} \mathsf{E}\,(\eta_n|F_{n-1}) &= \sum_{i\geq n} [\mathsf{E}\,(\mathsf{E}\,(Y_i|F_n)|F_{n-1}) - \mathsf{E}\,(Y_i|F_{n-1})] \\ &= \sum_{i\geq n} [\mathsf{E}\,(Y_i|F_{n-1}) - \mathsf{E}\,(Y_i|F_{n-1})] = 0. \end{aligned}$$

Decomposition for sums

With these preliminaries we now turn back to the sums, defined by (3.1) and (3.2). It turns out in this case that (η_k), $k \geq 1$, in (3.14) is a sequence of i.i.d random variables.

Lemma 3.2. Let the sequence $a(k)$, $k \geq 1$, satisfy the condition

$$\sum_{i=1}^{\infty} i|a(i)| < \infty.$$

Then S_n, defined by (3.2) and (3.4) admits the representations

$$S_n = \sum_{j=0}^{\infty} a(j) \sum_{k=1}^{n} \xi_k + (V_1 - V_{n+1}), \tag{3.15}$$

where (V_n), $n \geq 0$, is a strictly stationary sequence with

$$V_1 \stackrel{d}{=} \sum_{k=0}^{\infty} \sum_{j=k}^{\infty} a(j)\xi_k \tag{3.16}$$

($\stackrel{d}{=}$ denotes the equality in the distribution sence) such that for all $q > 0$

$$\limsup_{n\to\infty} n^{-1} \ln \mathsf{P}(|V_1| > nq) = -\infty. \tag{3.17}$$

Proof. To prove (3.15) we have first to check that under our condition, (3.13) holds for some $p \geq 2$. Pick $p = 2$. We show that

$$\sum_{k=0}^{\infty} (\mathsf{E}\,|\mathsf{E}\,(Y_k|F_0)|^2)^{1/2} < \infty.$$

Indeed, in view of (3.1) with $N = \infty$, we have

$$\begin{aligned}
\sum_{k=0}^{\infty} (\mathsf{E}\,|\mathsf{E}\,(Y_k|F_0)|^2)^{1/2} &= \sum_{k=0}^{\infty} \left(\mathsf{E} \left| \mathsf{E} \left(\sum_{j=-\infty}^{k} a(k-j)\xi_j | F_0 \right) \right|^2 \right)^{1/2} \\
&= \sum_{k=0}^{\infty} \left(\mathsf{E} \left(\sum_{j=-\infty}^{0} a(k-j)\xi_j \right)^2 \right)^{1/2} \\
&= \sum_{k=0}^{\infty} \left(\sum_{j=-\infty}^{0} a^2(k-j) \right)^{1/2} = \sum_{k=0}^{\infty} \left(\sum_{j=k}^{\infty} a^2(j) \right)^{1/2} \\
&\leq \sum_{k=0}^{\infty} \sum_{j=k}^{\infty} |a(j)| \leq \sum_{k=1}^{\infty} k|a(k)| < \infty.
\end{aligned}$$

Thus, we arrive at the conclusion that the desired decomposition exists. It is then an easy matter to calculate $M_n = \sum_{k=1}^{n} \eta_k$:

$$\begin{aligned}
M_n &= \sum_{k=1}^{n} \sum_{i \geq k} [\mathsf{E}\,(Y_i|F_k) - \mathsf{E}\,(Y_i|F_{k-1})] \\
&= \sum_{k=1}^{n} \sum_{i \geq k} \left[\mathsf{E} \left(\sum_{j=-\infty}^{i} a(i-j)\xi_j | F_k \right) - \mathsf{E} \left(\sum_{j=-\infty}^{i} a(i-j)\xi_j | F_{k-1} \right) \right] \\
&= \sum_{k=1}^{n} \sum_{i \geq k} \left[\sum_{j=-\infty}^{k} a(i-j)\xi_j - \sum_{j=-\infty}^{k-1} a(i-j)\xi_j \right] \\
&= \sum_{j=0}^{\infty} a(j) \sum_{k=1}^{n} \xi_k.
\end{aligned}$$

The similar calculations lead us to the following expression for V_n:

$$\begin{aligned} V_n &= \sum_{i\geq n} \mathsf{E}\,(Y_i|F_{n-1}) = \sum_{i\geq n} \mathsf{E}\left(\sum_{j=-\infty}^{i} a(i-j)\xi_j|F_{n-1}\right) \\ &= \sum_{i\geq n}\sum_{j=-\infty}^{n-1} a(i-j)\xi_j. \end{aligned}$$

Since (V_n), $n \geq 0$, is clearly a strictly stationary sequence, we have

$$V_1 \stackrel{d}{=} V_n \stackrel{d}{=} \sum_{k=0}^{\infty}\sum_{j=k}^{\infty} a(j)\xi_k.$$

It remains to prove (3.17). To this end, note that by Chebyshov's inequality with some $b > 0$ we have

$$\begin{aligned} \mathsf{P}(\pm V_1 > nq) &= \mathsf{P}\left(\pm\sum_{k=0}^{\infty}\sum_{j=k}^{\infty} a(j)\xi_k > nq\right) \\ &\leq \exp(-nb)\,\mathsf{E}\,\exp\left(\pm\frac{b}{q}\sum_{k=0}^{\infty}\sum_{j=k}^{\infty} a(j)\xi_k\right). \end{aligned} \tag{3.18}$$

We express now the right-hand side of the last inequality in terms of $G_\mu(\lambda) = \ln \mathsf{E}\,\exp(\lambda\xi_1)$:

$$\begin{aligned} \mathsf{E}\,\exp\left(\pm\frac{b}{q}\sum_{k=0}^{\infty}\sum_{j=k}^{\infty} a(j)\xi_k\right) &= \prod_{k=0}^{\infty} \mathsf{E}\,\exp\left(\pm\frac{b}{q}\sum_{j=k}^{\infty} a(j)\xi_k\right) \\ &= \exp\left(\sum_{k=0}^{\infty} G_\mu\left(\pm\frac{b}{q}\sum_{j=k}^{\infty} a(j)\right)\right). \end{aligned}$$

Note that under our condition

$$0 \leq \sum_{k=0}^{\infty} G_\mu\left(\pm\frac{b}{q}\sum_{j=k}^{\infty} a(j)\right) < \infty. \tag{3.19}$$

Denote

$$p_k = \pm\frac{b}{q}\sum_{j=k}^{\infty} a(j),$$

and

$$r = \sup_k |p_k|.$$

In virtue of assumption (3.4), $\lim_{k\to\infty} p_k = 0$ and hence, $r < \infty$. Thus, the following estimate is valid:

$$G_\mu(p_k) = \int_0^{p_k} G'_\mu(\lambda)\,d\lambda \leq \sup_{|\lambda|\leq r} |G'_\mu||p_k| \leq \frac{b}{q}\sup_{|\lambda|\leq r} |G'_\mu(\lambda)|\sum_{j=k}^{\infty} |a(j)|.$$

Finally, by (3.4), we obtain

$$\sum_{k=0}^{\infty} G_\mu(p_k) \le \frac{b}{q} \sup_{|\lambda|\le r} |G'_\mu| \sum_{k=1}^{\infty} k|a(k)| < \infty.$$

Combining the last estimate and (3.18) we get

$$n^{-1} \ln \mathsf{P}(\pm V_1 > nq) \le -b + n^{-1} \sum_{k=0}^{\infty} G_\mu \left(\pm \frac{b}{q} \sum_{j=k}^{\infty} a(j) \right).$$

Since b is arbitrary, we arrive at the statement of Lemma 3.2.

Large deviation principle in R

Having made these preparations we are now ready to prove the local large deviation principle in R.

Theorem 3.1. Let

$$\sum_{i=1}^{\infty} i|a(i)| < \infty$$

and

$$\sum_{k=0}^{\infty} a(k) > 0.$$

Then the sequence (S_n/n) obeys the local large deviation principle with the rate function

$$L\left(\frac{y}{\sum_{k=0}^{\infty} a(k)}\right),$$

$$L(x) = \sup_{\lambda\in R} (\lambda x - G_\mu(\lambda)),$$

i.e.,

$$\begin{aligned}
\limsup_{\delta\to 0} \limsup_{n} n^{-1} \ln \mathsf{P}\left(\left|\frac{S_n}{n} - y\right| \le \delta\right) \\
&= \liminf_{\delta\to 0} \liminf_{n} n^{-1} \ln \mathsf{P}\left(\left|\frac{S_n}{n} - y\right| \le \delta\right) \\
&= -L\left(\frac{y}{\sum_{k=0}^{\infty} a(k)}\right).
\end{aligned}$$

Proof. Theorem is proved by applying Lemma 3.1 and Lemma 3.2. To this end, we first show that (M_n/n) satisfies the local large deviation property with the rate function $L(y/\sum_{k=0}^{\infty} a(k))$. Indeed, by Lemma 3.2, M_n is the sum of i.i.d random variables with the logarithmic moment generating function

$$G(\lambda) = \ln \mathsf{E} \exp(\lambda\eta_1) = \ln \mathsf{E} \exp\left(\lambda \left(\sum_{j=0}^{\infty} a(j)\right) \xi_k\right),$$

$$G_\mu\left(\lambda\sum_{j=0}^{\infty}a(j)\right) < \infty.$$

Define $I(x)$ as the Legendre transformation of $G(\lambda)$:

$$I(x) = \sup_\lambda\,[\lambda x - G(\lambda)] = \sup_\lambda\left[\lambda x - G_\mu\left(\lambda\sum_{j=0}^{\infty}a(j)\right)\right] = L\left(\frac{y}{\sum\limits_{k=0}^{\infty}a(k)}\right).$$

Thus, by Cramér's theorem (S_n/n) obeys the full and local large deviation principle.

Hence, it remains to prove that $U_n = (V_1 - V_{n+1})$ satisfies (3.8). To this end, note that U_n is a strictly stationary sequence and

$$\mathsf{P}(|U_n| > nq) \le 2\mathsf{P}(|V_1| > nq).$$

Thus, by applying (3.17) we get the required result.

Remark 3.2. If $\sum_{k=0}^{\infty}a(k) < 0$ it is sufficient to consider

$$M_n = -\sum_{k=1}^{n}\left(\sum_{j=0}^{\infty}a(j)\right)(-\xi_k)$$

and take

$$I(y) = L\left(\frac{y}{-\sum\limits_{k=0}^{\infty}a(k)}\right)$$

instead of $L(y/\sum_{k=0}^{\infty}a(k))$.

If $\sum_{k=0}^{\infty}a(k) = 0$ it is the consequence of Lemma 3.2 that $S_n = V_1 - V_{n+1}$. Then, in view of (3.17) we obtain that $I(y) = \infty$.

Large deviations in function space

It is an easy step from the preceding to the function space version of the large deviation principle for the sums.

Keeping in mind the decomposition of Lemma 3.2, define $M^n = (M_t^n)$, $0 \le t \le 1$, with

$$M_t^n = n^{-1}\sum_{k=1}^{[nt]}\eta_k = n^{-1}\sum_{k=1}^{[nt]}\xi_k\sum_{j=0}^{\infty}a(j), \tag{3.20}$$

and

$$V_t^n = \sum_{i\ge[nt]+1}\ \sum_{j=-\infty}^{[nt]}a(i-j)\xi_i. \tag{3.21}$$

Then, by Lemma 3.2

$$W_t^n = M_t^n + U_t^n, \tag{3.22}$$

with

$$U_t^n = (V_1 - V_t)/n. \tag{3.23}$$

Clearly, nM_t^n is the sum of i.i.d random variables $\xi_k\sum_{j=0}^{\infty}a(j)$.

Suppose that $\sum_{j=0}^{\infty} a(j) > 0$. Then, by Theorem 1.13 in Section 1.5, M^n satisfies the local and full large deviation principle with rate function

$$I(\varphi) = \int_0^1 L\left(\frac{\dot{\varphi}(s)}{\sum_{j=0}^{\infty} a(j)}\right) \mathrm{d}s.$$

In view of Lemma 3.1, all that we have to do is to show that for every $q > 0$

$$\limsup_n n^{-1} \ln \mathsf{P}\left(\sup_{0\leq t\leq 1} |U_t^n| > q\right) = -\infty. \tag{3.24}$$

To this end, recalling that $V_1 \stackrel{d}{=} V_t$, note that

$$\mathsf{P}\left(\sup_{0\leq t\leq 1} |U_t^n| > q\right) \leq 2\mathsf{P}(|V_1| > q/2).$$

Then the argument used to prove the finite dimension result leads to the conclusion that (3.24) holds.

Thus we obtain the C-local large deviation principle for the sequence $W^n = (W_t^n)_{t\leq 1}$, $n \geq 1$.

Theorem 3.2. The sequence W^n obeys the C-local large deviation principle with the rate function

$$I(\varphi) = \int_0^1 L\left(\frac{\dot{\varphi}(s)}{\sum_{j=0}^{\infty} a(j)}\right) \mathrm{d}s.$$

3.2. LARGE DEVIATIONS FOR THE RECURSIVE SCHEME WITH THE WOLD-TYPE DISTURBANCES

Setting

In this section we study large deviations for a recursive scheme

$$Z_k^n - Z_{k-1}^n = n^{-1}(b_k(Z^{n,k-1}) + Y_k), \qquad Z_0^n = x, \tag{3.25}$$

$n \geq k \geq 1$ with Y_k, defined by (3.1) in Section 3.1, and $b_k(Z^{n,k-1})$, defined by (2.1) in Chapter 2.

Introduce a family of processes $Z^n = (Z_t^n)$, $0 \leq t \leq 1$, with

$$Z_t^n = x + \int_0^t b^n(Z_{[0,s)}^n)\, \mathrm{d}A_s^n + M_t^n + U_t^n, \tag{3.26}$$

where

$$W_t^n = M_t^n + U_t^n,$$

and

$$U_t^n = (V_1 - V_t)/n$$

are defined by (3.20)–(3.23) in the previous section and $b^n(Z^n_{[0,s)})$, defined by (2.2) in Chapter 2. Consider the sequence of processes $X^n = (X^n_t)$, $n \geq 1$, with

$$X^n_t = x + \int_0^t b^n(X^n_{[0,s)})\, dA^n_s + M^n_t. \tag{3.27}$$

Assume that b^n satisfies the conditions of Theorem 2.1 in Chapter 2, i.e.,

(i) for all $X \in D[0,1]$

$$\sup_n |b^n(X_{[0,t)})| \leq K < \infty,$$

(ii) for all $X, Y \in D[0,1]$

$$\sup_n |b^n(X_{[0,t)}) - b^n(Y_{[0,t)})| \leq K \sup_{0 \leq s \leq t} |X_s - Y_s|,$$

(iii) there exists a continuous mapping b: $C[0,1] \to R$ such that for each $\varphi \in C[0,1]$

$$\lim_n \int_0^1 |b^n(\varphi_{[0,t]}) - b(\varphi_{[0,t]})|\, dA^n_t = 0.$$

Lemma 3.3. The sequence X^n has the large deviation property with the rate function

$$I(\varphi) = \int_0^1 L\left(\frac{\dot\varphi(s) - b(\varphi_{[0,s]})}{\sum_{j=0}^{\infty} a(j)}\right) ds, \tag{3.28}$$

with

$$L(y) = \sup_\lambda (\lambda y - G_\mu(\lambda))$$

and $\varphi \in C[0,1]$, $\varphi(0) = x$.

Proof. The statement is the consequence of Theorem 2.1 in Chapter 2.

Denote

$$\Delta^n_t = Z^n_t - X^n_t,$$

and observe that from (3.26) and (3.27) we have

$$\Delta^n_t = \int_0^t [b^n(Z^n_{[0,s)}) - b^n(X^n_{[0,s)})]\, dA^n_s + U^n_t. \tag{3.29}$$

Keeping in mind condition (ii), let us introduce a random process $\alpha^n = (\alpha^n(s))$, $0 \leq s \leq 1$, with $\alpha^n_k = \alpha^n(k/n)$ such that

$$\sup_n |\alpha^n(s)| \leq K, \qquad 0 \leq s \leq 1,$$

and

$$\Delta_t^n = \int_0^t \Delta_{s-}^n \alpha^n(s-)\, dA_s^n + U_t^n. \tag{3.30}$$

Define

$$B_t^n = \int_0^t \alpha^n(s-)\, dA_s^n = n^{-1} \sum_{k=1}^{[nt]} \alpha_k^n, \tag{3.31}$$

and note that for sufficiently large n

$$|\Delta B_t^n| \leq |n^{-1}\alpha_k^n| < 1/2,$$

where

$$\begin{aligned} \Delta B_t^n &= B_t^n - B_{t-}^n, \\ B_{t-} &= \lim_{s \uparrow t} B_s, \qquad t > 0. \end{aligned}$$

Combining (3.30) and (3.31) we get

$$\Delta_t^n = \int_0^t \Delta_{s-}^n\, dB_s^n + U_t^n. \tag{3.32}$$

In view of Lemma 3.1 in Section 3.1 we conclude from the preceding that the C-local large deviation principle will be obtained once we show that for each $\delta > 0$,

$$\limsup_n n^{-1} \ln \mathrm{P}\left(\sup_{t \leq 1} |\Delta_t^n| > \delta \right) = -\infty. \tag{3.33}$$

Local large deviation principle

Before proving (3.33), we recall some definitions and results of stochastic calculus, adopted to our case. All these facts are well known and can be found in many text in which the modern theory of stochastic processes is discussed (see, for example, Liptser and Shiryaev (1989)).

We have already introduced the stochastic exponent and used it to define a Radon–Nikodym factor in changing the given measure. The discussion, outlined below, relies on the definition of the stochastic exponent as a solution of the linear Doleans-Dead equation

$$Y_t^n = 1 + \int_0^t Y_{s-}^n\, dB_s^n. \tag{3.34}$$

We will not discuss the definition of integral $\int_0^t Y_{s-}^n\, dB_s^n$, all the more in our case B_t^n is a function with a locally bounded variation and the integral is a Lebesgue–Stelties one. The equation (3.34) has a unique solution $Y_t^n \equiv \mathcal{E}_t^n$, named *the stochastic exponent*, which in the particular case admits the representation

$$Y_t^n \equiv \mathcal{E}_t^n = \exp(B_t^n) \prod_{s \leq t} (1 + \Delta B_s^n) \exp(-\Delta B_s^n) = \prod_{s \leq t} (1 + \Delta B_s^n). \tag{3.35}$$

We need the following formula integration by parts which is a consequence of the Itô formula:

$$\begin{aligned} X_tY_t &= X_0Y_0 + \int_0^t X_s\, dY_s + \int_0^t Y_{s-}\, dX_s \\ &= X_0Y_0 + \int_0^t X_{s-}\, dY_s + \int_0^t Y_s\, dX_s, \end{aligned} \tag{3.36}$$

where X, Y are functions with a locally bounded variation.

With these preliminaries taken care off, we are now ready to prove (3.33).

Lemma 3.4. Under our conditions, (3.33) holds.

Proof. We first show that

$$\Delta_t^n = \mathcal{E}_t^n \int_0^t (\mathcal{E}_s^n)^{-1}\, dU_s^n, \tag{3.37}$$

$\Delta_0^n = 0$. Indeed, denote

$$R_t^n = \int_0^t (\mathcal{E}_s^n)^{-1}\, dU_s^n,$$

$R_0^n = 0$. By applying the Itô formula (3.36) to $\mathcal{E}_t^n R_t^n$ and taking into account that by (3.34), $d\mathcal{E}_s^n = \mathcal{E}_{s-}^n\, dB_s^n$, we obtain:

$$\begin{aligned} \mathcal{E}_t^n R_t^n &= \int_0^t \mathcal{E}_s^n\, dR_s^n + \int_0^t R_{s-}^n\, d\mathcal{E}_s^n \\ &= \int_0^t \mathcal{E}_s^n (\mathcal{E}_s^n)^{-1}\, dU_s^n + \int_0^t R_{s-}^n \mathcal{E}_{s-}^n\, dB_s^n \\ &= U_t^n + \int_0^t \Delta_{s-}^n\, dB_s^n = \Delta_t^n. \end{aligned}$$

The second step is to show that

$$(\mathcal{E}_t^n)^{-1} = 1 - \int_0^t (\mathcal{E}_{s-}^n)^{-1} \frac{dB_s^n}{1 + \Delta B_s^n}. \tag{3.38}$$

Indeed, by (3.35),

$$(\mathcal{E}_t^n)^{-1} = \prod_{s \le t} \frac{1}{(1 + \Delta B_s^n)} = \prod_{s \le t} \left(1 - \frac{\Delta B_s^n}{(1 + \Delta B_s^n)}\right). \tag{3.39}$$

Denoting

$$D_t^n = -\int_0^t \frac{dB_s^n}{1 + \Delta B_s^n},$$

with

$$\Delta D^n_s = -\frac{\Delta B^n_s}{1+\Delta B^n_s},$$

we write (3.39) in equivalent form (see (3.35)):

$$(\mathcal{E}^n_t)^{-1} = \exp(-D^n_t)\prod_{s\le t}(1+\Delta D^n_s)\exp(-\Delta D^n_s).$$

It is then an easy step to the conclusion that $(\mathcal{E}^n_t)^{-1}$ is the solution of the following Doleans-Dead equation:

$$(\mathcal{E}^n_t)^{-1} = 1 - \int_0^t (\mathcal{E}^n_{s-})^{-1}\,\mathrm{d}D^n_s.$$

This gives (3.38).

We next apply (3.36) to $(\mathcal{E}^n_t)^{-1}U^n_t$ to get, taking into account (3.38), that

$$(\mathcal{E}^n_t)^{-1}U^n_t = \int_0^t U^n_{s-}(\mathcal{E}^n_{s-})^{-1}\frac{\mathrm{d}B^n_s}{1+\Delta B^n_s} + \int_0^t (\mathcal{E}^n_s)^{-1}\,\mathrm{d}U^n_s. \tag{3.40}$$

Turning to the estimation of $\sup_{t\le 1}|\Delta_t|$, we note that from (3.37)

$$\sup_{t\le 1}|\Delta^n_t| \le \sup_{t\le 1}\mathcal{E}^n_t \sup_{t\le 1}\left|\int_0^t (\mathcal{E}^n_s)^{-1}\,\mathrm{d}U^n_s\right|. \tag{3.41}$$

From (3.40) we obtain the following estimate:

$$\begin{aligned}\sup_{t\le 1}\left|\int_0^t (\mathcal{E}^n_s)^{-1}\,\mathrm{d}U^n_s\right| \le{}& \sup_{t\le 1}(\mathcal{E}^n_t)^{-1}\sup_{t\le 1}|U^n_t| \\ &+2\sup_{t\le 1}|U^n_t|\sup_{t\le 1}(\mathcal{E}^n_t)^{-1}\sum_{t\le 1}|\Delta B^n_t|. \end{aligned} \tag{3.42}$$

Note that by definition (3.31), for sufficiently large n

$$|\Delta B^n_t| < 1/2$$

and hence

$$\sum_{t\le 1}|\Delta B^n_t| \le \text{const}.$$

Thus, from (3.42)

$$\sup_{t\le 1}\left|\int_0^t (\mathcal{E}^n_s)^{-1}\,\mathrm{d}U^n_s\right| \le C\sup_{t\le 1}(\mathcal{E}^n_t)^{-1}\sup_{t\le 1}|U^n_t|.$$

Tying together (3.41) and (3.42) we get

$$\sup_{t\le 1}|\Delta^n_t| \le C\sup_{t\le 1}\mathcal{E}^n_t\sup_{t\le 1}(\mathcal{E}^n_t)^{-1}\sup_{t\le 1}|U^n_t|.$$

Since by (3.24), for each $q > 0$,

$$\lim_n n^{-1} \ln \mathsf{P}\left(\sup_{0\le t\le 1} |U_t^n| > q\right) = -\infty,$$

it remains to estimate $\sup_{t\le 1}(\mathcal{E}_t^n)^{-1}$. The required estimate comes from (3.39) and the fact that for sufficiently large n, $|\Delta B_t^n| < 1/2$:

$$\begin{aligned}(\mathcal{E}_t^n)^{-1} &= \prod_{s\le t}\left(1 - \frac{\Delta B_s^n}{(1+\Delta B_s^n)}\right) = \exp\left(\sum_{s\le t} \ln\left[1 - \frac{\Delta B_s^n}{(1+\Delta B_s^n)}\right]\right)\\ &\le \exp\left(\sum_{s\le t} \ln\left[1 - 2\Delta B_s^n\right]\right) \le \mathrm{e}^{-C}.\end{aligned}$$

Lemma 3.4 is proved.

By combining Lemma 3.1 with Lemma 3.4 we arrive at the C-local large deviation principle.

Theorem 3.3. The sequence Z^n obeys the C-local large deviation principle with the rate function, defined by (2.4).

Comments

The results of this chapter are obtained by Gulinsky and Liptser, and as far as we know are new.

Chapter 4

Generalization of Cramér's theorem

4.1. LARGE DEVIATIONS FOR SUMS OF STATIONARY SEQUENCES

Setting. Main results

Let $X = (X_k)$, $-\infty < k < \infty$, be a strong stationary ergodic sequence, defined on (Ω, F, P) with values in R and $\mathsf{E}X_1 = 0$. Define a filtration $\mathbf{F} = (F_k)$, $k \geq 0$, with

$$F_k = \sigma(X_i, -\infty < i \leq k).$$

Assume that X satisfies Cramér's condition for all $\lambda \in R$, i.e.,

$$\mathsf{E}e^{\lambda X_1} < \infty.$$

Define the sequence $S = (S_n)$, $n \geq 1$, with

$$S_n = \sum_{k=1}^{n} X_k. \tag{4.1}$$

By the ergodic property

$$n^{-1} S_n \to 0.$$

Then S is a candidate for the large deviation theory.

Define the stochastic exponent

$$\mathcal{E}_n(\lambda) = \prod_{k=1}^{n} \mathsf{E}\,(e^{\lambda X_k} | F_{k-1}), \tag{4.2}$$

and the stochastic cummulant

$$\ln \mathcal{E}_n(\lambda) = \sum_{k=1}^{n} \ln \mathsf{E}\,(e^{\lambda X_k} | F_{k-1}). \tag{4.3}$$

Evidently, the cummulant is a strong stationary ergodic sequence and by Birkhoff and Khinchin's ergodic theorem there exists a function $G(\lambda)$ such that

$$\lim_{n\to\infty} n^{-1} \sum_{k=1}^{n} \ln \mathsf{E}\,(e^{\lambda X_k} | F_{k-1}) = G(\lambda). \tag{4.4}$$

with

$$G(\lambda) = \mathsf{E}\,(\ln \mathsf{E}\,(e^{\lambda X_1} | F_0)). \tag{4.5}$$

Assume that

$$\lim_{|\lambda|\to\infty} \frac{G(\lambda)}{|\lambda|} = \infty. \tag{R}$$

The stochastic cummulant and $G(\lambda)$ are convex functions. Define

$$L(x) = \sup_{\lambda \in R} [\lambda x - G(\lambda)].$$

The problem arises to find a condition such that $L(x)$ is the rate function in the large deviation problem for the sequence (S_n/n), $n \geq 1$.

It turns out that $L(x)$ is in fact the rate function in this large deviation problem, if $n^{-1}\sum_{k=1}^{n} \ln \mathsf{E}(e^{\lambda X_k}|F_{k-1})$ converges to $G(\lambda)$ 'exponentially fast', i.e., for every $\varepsilon > 0$

$$\lim_n n^{-1} \ln \mathsf{P}\left(\left|n^{-1}\sum_{k=1}^{n} \ln \mathsf{E}(e^{\lambda X_k}|F_{k-1}) - G(\lambda)\right| > \varepsilon\right) = -\infty. \tag{4.6}$$

Theorem 4.1. Assume that condition (4.6) holds. Then under Cramér's condition the sequence (S_n/n), $n \geq 1$, has the large deviation property with the rate function $L(x)$.

Remark 4.1. If one considers the proof, given below, one can recognize that the result remains valid for any (not necessary stationary) sequence X which meets (4.6), provided $G(\lambda)$ satisfies some regularity conditions.

For proving Theorem 4.1 we will once again use a transform of the given measure. Define

$$z_n(\lambda) = \prod_{k=1}^{n} \frac{e^{\lambda X_k}}{\mathsf{E}(e^{\lambda X_k}|F_{k-1})} = \exp(\lambda S_n - \ln \mathcal{E}_n(\lambda)). \tag{4.7}$$

$(z_k(\lambda), F_k)$ is a martingale and $\mathsf{E} z_n(\lambda) = 1$. Define a new measure $Q^{n,\lambda}$ by

$$dQ^{n,\lambda} = z_n(\lambda)\, d\mathsf{P}. \tag{4.8}$$

Ergodic property S_n w.r.t. $Q^{n,\lambda}$

We first present two lemmas and then tie them together to obtain the ergodic property. Denote

$$\alpha_k(\lambda) = \frac{e^{\lambda X_k}}{\mathsf{E}(e^{\lambda X_k}|F_{k-1})}, \tag{4.9}$$

and define

$$\begin{aligned}\sum_{k=1}^{n} \ln \mathsf{E}(\alpha_k^2(\lambda)|F_{k-1}) &= \sum_{k=1}^{n} \ln \mathsf{E}(e^{2\lambda X_k}|F_{k-1}) - 2\sum_{k=1}^{n} \ln \mathsf{E}(e^{\lambda X_k}|F_{k-1}) \\ &= \ln \mathcal{E}_n(2\lambda) - 2\ln \mathcal{E}_n(\lambda).\end{aligned} \tag{4.10}$$

By condition (4.6), we have

$$\lim_n \mathsf{P}\left(\left|n^{-1}\sum_{k=1}^{n} \ln \mathsf{E}(\alpha_n^2(\lambda)|F_{k-1}) - [G(2\lambda) - 2G(\lambda)]\right| > \varepsilon\right) = -\infty. \tag{4.11}$$

Introduce a stopping time

$$\tau_n(\lambda) = \left(\min j:\ \sum_{k=1}^{j+1} \ln \mathsf{E}\,(\alpha_k^2(\lambda)|F_{k-1}) > 2n[G(2\lambda) - 2G(\lambda)]\right). \tag{4.12}$$

Lemma 4.1. Let (4.11) holds. Then

$$\lim_n n^{-1} \ln \mathsf{P}(\tau_n(\lambda) < n) = -\infty, \tag{4.13}$$

and

$$\lim_n n^{-1} \ln Q^{n,\lambda}(\tau_n(\lambda) < n) = -\infty. \tag{4.14}$$

Proof. We first note that

$$\sum_{k=1}^{n} \ln \mathsf{E}\,(\alpha_k^2(\lambda)|F_{k-1}), \qquad n \geq 1,$$

is a non-decreasing sequence. Indeed, since $\alpha_k^2(\lambda)$ is a convex function, by Jensen's inequality, we have

$$\mathsf{E}\,(\alpha_k^2(\lambda)|F_{k-1}) \geq (\mathsf{E}\,(\alpha_k(\lambda)|F_{k-1}))^2 = 1.$$

By definition of $\tau_n(\lambda)$, taking into account the last inequality, we get

$$\begin{aligned}(\tau_n(\lambda) < n) &= (\tau_n(\lambda) + 1 < n + 1)\\ &\subseteq \left(\sum_{k=1}^{n+1} \ln \mathsf{E}\,(\alpha_k^2(\lambda)|F_{k-1}) \geq \sum_{k=1}^{\tau_n(\lambda)+1} \ln \mathsf{E}\,(\alpha_k^2(\lambda)|F_{k-1})\right)\\ &\subseteq \left(\sum_{k=1}^{n+1} \ln \mathsf{E}\,(\alpha_k^2(\lambda)|F_{k-1}) \geq 2n[G(2\lambda) - 2G(\lambda)]\right).\end{aligned} \tag{4.15}$$

Thus, from (4.11), we obtain (4.13).

Turning to the proof (4.14), we observe that $(\tau_n(\lambda) < n)$ is F_n and $F_{\tau_n(\lambda)}$-measurable. Thus,

$$\begin{aligned}Q^{n,\lambda}(\tau_n(\lambda) < n) &= \mathsf{E}\,\mathrm{I}\,(\tau_n(\lambda) < n)\, z_n(\lambda)\\ &= \mathsf{E}\,(\mathsf{E}\,(\mathrm{I}\,(\tau_n(\lambda) < n)\, z_n(\lambda)|F_{\tau_n(\lambda)}))\\ &= \mathsf{E}\,\mathrm{I}\,(\tau_n(\lambda) < n)\,\mathsf{E}\,(z_n(\lambda)|F_{\tau_n(\lambda)}).\end{aligned} \tag{4.16}$$

By martingale property,

$$\mathsf{E}\,(z_n(\lambda)|F_{n\wedge\tau_n(\lambda)}) = z_{n\wedge\tau_n(\lambda)}(\lambda),$$

and so from (4.16), we get

$$Q^{n,\lambda}(\tau_n(\lambda) < n) = \mathsf{E}\,\mathrm{I}\,(\tau_n(\lambda) < n) z_{n\wedge\tau_n(\lambda)}(\lambda). \tag{4.17}$$

By Cauchy and Bunjakovskii's inequality

$$Q^{n,\lambda}(\tau_n(\lambda) < n) \le (\mathsf{P}(\tau_n(\lambda) < n))^{1/2}(\mathsf{E}\,(z_{n\wedge\tau_n(\lambda)}(\lambda))^2)^{1/2}. \tag{4.18}$$

We next write $(z_{n\wedge\tau_n(\lambda)}(\lambda))^2$ in the following form:

$$\begin{aligned}(z_{n\wedge\tau_n(\lambda)}(\lambda))^2 &= \exp(2\lambda S_{n\wedge\tau_n(\lambda)} - 2\ln\mathcal{E}_{n\wedge\tau_n(\lambda)}(\lambda))\\ &= \exp(2\lambda S_{n\wedge\tau_n(\lambda)} - \ln\mathcal{E}_{n\wedge\tau_n(\lambda)}(2\lambda)\\ &\quad + \ln\mathcal{E}_{n\wedge\tau_n(\lambda)}(2\lambda) - 2\ln\mathcal{E}_{n\wedge\tau_n(\lambda)}(\lambda))\\ &= z_{n\wedge\tau_n(\lambda)}(2\lambda)\exp\left(\sum_{k=1}^{\tau_n(\lambda)\wedge n}\ln\mathsf{E}\,(\alpha_k^2(\lambda)|F_{k-1})\right).\end{aligned}$$

By definition (4.12) of $\tau_n(\lambda)$, from the last equality, we obtain

$$(z_{n\wedge\tau_n(\lambda)}(\lambda))^2 \le z_{n\wedge\tau_n(\lambda)}(2\lambda)\,\exp(2n[G(2\lambda) - 2G(\lambda)]),$$

and since by martingale property, $\mathsf{E}z_{n\wedge\tau_n(2\lambda)}(\lambda) = 1$, we get

$$\mathsf{E}\,(z_{n\wedge\tau_n(\lambda)}(\lambda))^2 \le \exp(2n[G(2\lambda) - 2G(\lambda)]). \tag{4.19}$$

Finally, by combining (4.19) with (4.18), we arrive at the following estimate:

$$\varlimsup_n n^{-1}\ln Q^{n,\lambda}(\tau_n(\lambda) < n) \le \varlimsup_n \frac{1}{2n}\ln\mathsf{P}(\tau_n(\lambda) < n) + G(2\lambda) - 2G(\lambda).$$

Thus, by (4.13) we get (4.14). This completes the proof of Lemma 4.1.

Lemma 4.2. Let (A_n), $n \ge 1$, be a sequence of subsets in Ω with the property

$$\varlimsup_n n^{-1}\ln\mathsf{P}(A_n) = -\infty. \tag{4.20}$$

Assume that (4.11) holds. Then

$$\varlimsup_n n^{-1}\ln Q^{n,\lambda}(A_n) = -\infty.$$

Proof. We first note that

$$\begin{aligned}Q^{n,\lambda}(A_n) &= Q^{n,\lambda}(A_n,\tau_n(\lambda) < n) + Q^{n,\lambda}(A_n,\tau_n(\lambda) \ge n)\\ &\le Q^{n,\lambda}(\tau_n(\lambda) < n) + Q^{n,\lambda}(A_n,\tau_n(\lambda) \ge n)\\ &\le 2\max(Q^{n,\lambda}(\tau_n(\lambda) < n), Q^{n,\lambda}(A_n,\tau_n(\lambda) \ge n)).\end{aligned} \tag{4.21}$$

By (4.8), and since $n \wedge \tau_n(\lambda) = n$ on the set $(\tau_n(\lambda) \ge n)$, we have

$$\begin{aligned}Q^{n,\lambda}(A_n,\tau_n(\lambda) \ge n) &= \mathsf{E}\,\mathrm{I}(A_n,\tau_n(\lambda) \ge n)\,z_n(\lambda)\\ &= \mathsf{E}\,\mathrm{I}(A_n,\tau_n(\lambda) \ge n)\,z_{n\wedge\tau}(\lambda).\end{aligned}$$

By Cauchy and Bunjakovskii's inequality,

$$Q^{n,\lambda}(A_n, \tau_n(\lambda) \geq n) \leq (\mathsf{P}(A_n, \tau_n(\lambda) \geq n))^{1/2}(\mathsf{E}\,[z_{n\wedge\tau}(\lambda)]^2)^{1/2},$$

and so, taking into account (4.19), we get

$$n^{-1} \ln Q^{n,\lambda}(A_n, \tau_n(\lambda) \geq n) \leq \frac{1}{2n} \ln \mathsf{P}(A_n, \tau_n(\lambda) \geq n) + G(2\lambda) - 2G(\lambda).$$

Thus, by condition (4.20),

$$\lim_n n^{-1} \ln Q^{n,\lambda}(A_n, \tau_n(\lambda) \geq n) = -\infty. \tag{4.22}$$

Finally, by combining (4.21) with (4.14) and (4.22) we arrive at the conclusion that

$$\lim_n n^{-1} \ln Q^{n,\lambda}(A_n) \leq -\infty.$$

Lemma 4.2 is proved.

Remark 4.1. This result is closely related to the concept of contiguity of probability measures (see, for example, Liptser et al. (1982))

In our definition of $Q^{n,\lambda}$ the parameter λ is not yet fixed. From the regularity assumption (R) it follows that the supremum

$$L(x) = \sup_{\lambda \in R} [\lambda x - G(\lambda)]$$

is attained at some point λ^0 for the given arbitrary x. Pick λ^0 and note that

$$G'(\lambda^0) = x.$$

The following ergodic result clarifies in the present setting the idea of the measure transform.

Lemma 4.3. Let condition (4.6) hold. Then

$$\lim_n Q^{n,\lambda^0}\left(\left|\frac{S_n}{n} - x\right| > \delta\right) = 0. \tag{4.23}$$

Proof. Define (E^{n,λ^0} is the expectation w.r.t. Q^{n,λ^0})

$$z_n(\lambda^0, \mu) = \prod_{k=1}^{n} \frac{e^{\mu X_k}}{\mathsf{E}^{n,\lambda^0}(e^{\mu X_k} | F_{k-1})}, \tag{4.24}$$

and note that

$$\mathsf{E}^{n,\lambda^0} z_n(\lambda^0, \mu) = 1. \tag{4.25}$$

Using (4.9), we next represent $\mathsf{E}^{n,\lambda^0}(e^{\mu X_k}|F_{k-1})$ as the conditional expectation w.r.t P (see Chapter 2):

$$\begin{aligned}\mathsf{E}^{n,\lambda^0}(e^{\mu X_k}|F_{k-1}) &= \mathsf{E}\,(\alpha_k(\lambda^0)\, e^{\mu X_k}|F_{k-1}) \\ &= \frac{\mathsf{E}\,(e^{\lambda^0 X_k}\, e^{\mu X_k}|F_{k-1})}{\mathsf{E}\,(e^{\lambda^0 X_k}|F_{k-1})} = \frac{\mathsf{E}\,(e^{(\lambda^0+\mu) X_k}|F_{k-1})}{\mathsf{E}\,(e^{\lambda^0 X_k}|F_{k-1})}.\end{aligned}$$

Hence, from (4.24) and (4.3) we get

$$\begin{aligned} z_n(\lambda^0, \mu) &= \prod_{k=1}^{n} \frac{e^{\mu X_k} \mathsf{E}(e^{\lambda^0 X_k} | F_{k-1})}{\mathsf{E}(e^{(\lambda^0+\mu)X_k} | F_{k-1})} \\ &= \exp(\mu S_n - \ln \mathcal{E}_n(\lambda^0 + \mu) + \ln \mathcal{E}_n(\lambda^0)). \end{aligned} \tag{4.26}$$

Denote

$$A_n^\varepsilon(\lambda^0) = (|n^{-1} \ln \mathcal{E}_n(\lambda^0) - G(\lambda^0)| > \varepsilon),$$
$$A_n^\varepsilon(\lambda^0 + \mu) = (|n^{-1} \ln \mathcal{E}_n(\lambda^0 + \mu) - G(\lambda^0 + \mu)| > \varepsilon),$$

and

$$\bar{A}_n^\varepsilon(\lambda^0) = \Omega \backslash A_n^\varepsilon(\lambda^0), \bar{A}_n^\varepsilon(\lambda^0 + \mu) = \Omega \backslash A_n^\varepsilon(\lambda^0 + \mu).$$

Turning to prove (4.23) we note that it is sufficient to consider $\mathsf{P}(S_n/n > x + \delta)$. The set $(S_n/n < x - \delta)$ is treated similarly. To this end, note that from (4.25) and (4.26) we have:

$$\begin{aligned} 1 &\geq \mathsf{E}^{n,\lambda^0} \mathrm{I}\left(\frac{S_n}{n} > x + \delta, \bar{A}_n^\varepsilon(\lambda^0), \bar{A}_n^\varepsilon(\lambda^0 + \mu)\right) \\ &\quad \times \exp(\mu S_n - \ln \mathcal{E}_n(\lambda^0 + \mu) + \ln \mathcal{E}_n(\lambda^0)) \\ &\geq \exp(n\mu(x + \delta) - nG(\lambda^0 + \mu) - n\varepsilon + nG(\lambda^0) - n\varepsilon) \\ &\quad \times Q^{n,\lambda^0}\left(\frac{S_n}{n} > x + \delta, \bar{A}_n^\varepsilon(\lambda^0), \bar{A}_n^\varepsilon(\lambda^0 + \mu)\right). \end{aligned}$$

Thus,

$$Q^{n,\lambda^0}\left(\frac{S_n}{n} > x + \delta, \bar{A}_n^\varepsilon(\lambda^0), \bar{A}_n^\varepsilon(\lambda^0 + \mu)\right) \leq \exp(-n[\mu(x+\delta) - G(\lambda^0+\mu) + G(\lambda^0) - 2\varepsilon]).$$

Since $G(\lambda)$ is a smooth function and by the choice of λ^0 (recall that $G'(\lambda^0) = x$), for some θ, $0 < \theta < 1$, we have

$$\begin{aligned} &Q^{n,\lambda^0}\left(\frac{S_n}{n} > x + \delta, \bar{A}_n^\varepsilon(\lambda^0), \bar{A}_n^\varepsilon(\lambda^0 + \mu)\right) \\ &\quad \leq \exp\left(-n\left[\mu(x - G'(\lambda^0)) + \mu\delta - \frac{\mu^2}{2} G''(\lambda^0 + \theta\mu) - 2\varepsilon\right]\right) \\ &\quad = \exp\left(-\left[\mu\delta - \frac{\mu^2}{2} G''(\lambda^0 + \theta\mu) - 2\varepsilon\right]\right). \end{aligned}$$

Since $\varepsilon > 0$ and $\delta > 0$ are arbitrary, we can choose $\mu > 0$ such that

$$\mu\delta - \frac{\mu^2}{2} G''(\lambda^0 + \theta\mu) - 2\varepsilon > 0,$$

and so

$$Q^{n,\lambda^0}\left(\frac{S_n}{n} > x + \delta, \bar{A}_n^\varepsilon(\lambda^0), \bar{A}_n^\varepsilon(\lambda^0 + \mu)\right) \to 0, \tag{4.27}$$

as $n \to \infty$. The passage from (4.27) to (4.23) is accomplished with the help of the following inequality:

$$\begin{aligned} Q^{n,\lambda^0}\left(\frac{S_n}{n} > x + \delta\right) &\le Q^{n,\lambda^0}\left(\frac{S_n}{n} > x + \delta, \bar{A}_n^\varepsilon(\lambda^0), \bar{A}_n^\varepsilon(\lambda^0 + \mu)\right) \\ &\quad + Q^{n,\lambda^0}(A_n^\varepsilon(\lambda^0)) + Q^{n,\lambda^0}(A_n^\varepsilon(\lambda^0 + \mu)) \\ &\le 3\max\Big(Q^{n,\lambda^0}\left(\frac{S_n}{n} > x + \delta, \bar{A}_n^\varepsilon(\lambda^0), \bar{A}_n^\varepsilon(\lambda^0 + \mu)\right), \\ &\qquad Q^{n,\lambda^0}(A_n^\varepsilon(\lambda^0)), Q^{n,\lambda^0}(A_n^\varepsilon(\lambda^0 + \mu))\Big). \end{aligned}$$

Since by Lemma 4.2

$$\lim_n n^{-1} \ln Q^{n,\lambda^0}(A_n^\varepsilon(\lambda^0)) = -\infty,$$

and

$$\lim_n n^{-1} \ln Q^{n,\lambda^0}(A_n^\varepsilon(\lambda^0 + \mu)) = -\infty,$$

and in virtue of (4.27), we arrive at the statement of Lemma 4.3.

Lower bound

We have, at last, made all the preparations necessary to prove the lower bound.

Theorem 4.2. Under conditions of Theorem 4.1

$$\liminf_{\delta \to 0} \liminf_n \mathsf{P}\left(\left|\frac{S_n}{n} - x\right| \le \delta\right) \ge -L(x).$$

Proof. Since by definition $z_n(\lambda^0) > 0$, we have

$$\begin{aligned} \mathsf{P}\left(\left|\frac{S_n}{n} - x\right| \le \delta, \bar{A}_n^\varepsilon(\lambda^0)\right) &= \mathsf{E}^{n,\lambda^0} \mathrm{I}\left(\left|\frac{S_n}{n} - x\right| \le \delta, \bar{A}_n^\varepsilon(\lambda^0)\right)(z_n(\lambda^0))^{-1} \\ &= \mathsf{E}^{n,\lambda^0} \mathrm{I}\left(\left|\frac{S_n}{n} - x\right| \le \delta, \bar{A}_n^\varepsilon(\lambda^0)\right) \exp(\lambda S_n - \ln \mathcal{E}_n(\lambda)) \\ &\ge \mathsf{E}^{n,\lambda^0} \mathrm{I}\left(\left|\frac{S_n}{n} - x\right| \le \delta, \bar{A}_n^\varepsilon(\lambda^0\right) \exp(-\lambda^0 x + nG(\lambda^0) - n|\lambda^0|\delta - n\varepsilon) \\ &= \exp(-n[L(x) + |\lambda^0|\delta + \varepsilon])\, Q^{n,\lambda^0}\left(\left|\frac{S_n}{n} - x\right| \le \delta, \bar{A}_n^\varepsilon(\lambda^0)\right). \end{aligned} \tag{4.28}$$

We next note that

$$Q^{n,\lambda^0}\left(\left|\frac{S_n}{n} - x\right| \le \delta, \bar{A}_n^\varepsilon(\lambda^0)\right) \ge 1 - Q^{n,\lambda^0}\left(\left|\frac{S_n}{n} - x\right| > \delta\right) - Q^{n,\lambda^0}(A_n^\varepsilon(\lambda^0)),$$

and so by (4.23) of Lemma 4.3 and by Lemma 4.2, for each $\varepsilon > 0$, we get

$$\lim_n n^{-1} Q^{n,\lambda^0}\left(\left|\frac{S_n}{n} - x\right| \le \delta, \bar{A}_n^\varepsilon(\lambda^0)\right) = 1. \tag{4.29}$$

Since $\delta > 0$ and $\varepsilon > 0$ are arbitrary, by combining (4.29) with (4.28), we obtain

$$\begin{aligned}\liminf_{\delta\to 0}\liminf_n \mathsf{P}\left(\left|\frac{S_n}{n}-x\right|\le\delta\right) &\ge \liminf_{\delta\to 0}\liminf_n \mathsf{P}\left(\left|\frac{S_n}{n}-x\right|\le\delta, \bar{A}_n^\varepsilon(\lambda^0)\right)\\ &\ge -L(x).\end{aligned}$$

Theorem 4.2 is proved.

Upper bound
We will now prove a slightly deficient form of upper bound under weaker than assumption (4.6). Assume that there exists a lower semicontinuous function $J = J(\lambda)$, $J(\lambda) \ge 0$, $\lambda \in R$, such that for every $\varepsilon > 0$,

$$\limsup_n n^{-1}\ln \mathsf{P}(|n^{-1}\ln \mathcal{E}_n(\lambda) - G(\lambda)| > \varepsilon) \le -J(\lambda). \tag{4.30}$$

As in the proof of Cramér's theorem, it is sufficient to consider sets $(S_n/n \ge x)$ or $(|S_n/n - x| \le \delta)$.

Theorem 4.3. Let (4.30) holds. Then for every $\delta > 0$,

$$\limsup_n n^{-1}\ln \mathsf{P}\left(\left|\frac{S_n}{n}-x\right|\le\delta\right) \le -\sup_{\lambda\in R}([\lambda x - G(\lambda)] \wedge J(\lambda)),$$

and

$$\limsup_n n^{-1}\ln \mathsf{P}\left(\frac{S_n}{n}\ge x\right) \le -\sup_{\lambda\in R}([\lambda x - G(\lambda)] \wedge J(\lambda)).$$

Remark 4.2. If $\lambda^0 = \arg\sup_\lambda [\lambda x - G(\lambda)]$ and $J(\lambda^0) > I(x)$, then

$$\limsup_n n^{-1}\ln \mathsf{P}\left(\left|\frac{S_n}{n}-x\right|\le\delta\right) \le -I(x). \tag{4.31}$$

If $J(\lambda) = \infty$, (4.31) holds, evidently, for all x.

Proof. Let $A_n^\varepsilon(\lambda)$ and $\bar{A}_n^\varepsilon(\lambda)$ be the sets, defined as in the preceding paragraph. Note that

$$\begin{aligned}\mathsf{P}\left(\frac{S_n}{n}\ge x\right) &\le \mathsf{P}\left(\frac{S_n}{n}\ge x, \bar{A}_n^\varepsilon(\lambda)\right) + \mathsf{P}(A_n^\varepsilon(\lambda))\\ &\le 2\max\left[\mathsf{P}\left(\frac{S_n}{n}\ge x, \bar{A}_n^\varepsilon(\lambda)\right), \mathsf{P}(A_n^\varepsilon(\lambda))\right].\end{aligned}$$

This implies that for every $\varepsilon > 0$,

$$\begin{aligned}&\limsup_n n^{-1}\ln \mathsf{P}\left(\frac{S_n}{n}\ge x\right)\\ &\le \max\left[\limsup_n n^{-1}\ln \mathsf{P}\left(\frac{S_n}{n}\ge x, \bar{A}_n^\varepsilon(\lambda)\right), \limsup_n n^{-1}\ln \mathsf{P}(A_n^\varepsilon(\lambda))\right].\end{aligned} \tag{4.32}$$

Recall that $\mathsf{E}z_n(\lambda) = 1$, $n \geq 1$. Therefore,

$$1 \geq \mathsf{E}z_n(\lambda)\, \mathrm{I}\left(\frac{S_n}{n} \geq x, \bar{A}_n^\varepsilon(\lambda)\right). \tag{4.33}$$

On the set $\left(\frac{S_n}{n} \geq x, \bar{A}_n^\varepsilon(\lambda)\right)$ we have

$$\begin{aligned} z_n(\lambda) &= \exp\left(n\left[\lambda\frac{S_n}{n} - n^{-1}\ln \mathcal{E}_n(\lambda)\right]\right) \\ &\geq \exp\left(n[\lambda x - G(\lambda) - \varepsilon]\right). \end{aligned}$$

The last inequality and (4.33) imply that

$$\limsup_n n^{-1} \ln \mathsf{P}\left(\frac{S_n}{n} \geq x, \bar{A}_n^\varepsilon(\lambda)\right) \leq -[\lambda x - G(\lambda) - \varepsilon]. \tag{4.34}$$

Since ε is arbitrary, tying (4.34) and (4.32) together, we arrive at the following family ($\lambda \in R$) of inequalities:

$$\limsup_n n^{-1} \ln \mathsf{P}\left(\frac{S_n}{n} \geq x\right) \leq -([\lambda x - G(\lambda)] \wedge J(\lambda)).$$

By optimizing over this family, we get the required result. Theorem 4.3 is proved.

4.2. LARGE DEVIATIONS FOR SUMS OF SEMIMARTINGALES

Setting. Main results. Example

Let $X = (X_k)$, $k \geq 1$, be a sequence of random variables with values in R, defined on (Ω, F, P) with a filtration $\mathbf{F} = (F_k)$, $k \geq 0$. Assume that the sequence $S = (S_n)$, $n \geq 1$, with $S_n = \sum_{k=1}^n X_k$ has the ergodic property. We assume again that X satisfies Cramér's condition for all $\lambda \in R$, i.e.,

$$\mathsf{E}e^{\lambda X_k} \leq \infty, \qquad k \geq 1,$$

and define the stochastic exponent

$$\mathcal{E}_n(\lambda) = \prod_{k=1}^n \mathsf{E}\,(e^{\lambda X_k} | F_{k-1}),$$

and the stochastic cummulant

$$\ln \mathcal{E}_n(\lambda) = \sum_{k-1}^n \ln \mathsf{E}\,(e^{\lambda X_k} | F_{k-1}).$$

In the previous section we proved the large deviation property for sums of ergodic sequences under the strong condition (4.6). In this section we will formulate in terms of stochastic exponent a number of conditions which imply the large deviation principle for more general types of ergodic sequences and naturally with a different rate function.

Define

$$z_n(\lambda) = \prod_{k=1}^{n} \frac{e^{\lambda X_k}}{E(e^{\lambda X_k}|F_{k-1})} = e^{\lambda S_n}\mathcal{E}_n^{-1}(\lambda)$$

and

$$\tilde{z}_n(\lambda) = E(z_n(\lambda)|S_n) = \exp(\lambda S_n + \ln E(\mathcal{E}_n^{-1}(\lambda)|S_n)). \tag{4.35}$$

Note that $E\tilde{z}_n(\lambda) = Ez_n(\lambda) = 1$ and introduce a new measure

$$dQ^{n,\lambda} = \tilde{z}_n(\lambda)\,dP.$$

Since $z_n(\lambda) > 0$, we have

$$dP = (\tilde{z}_n(\lambda))^{-1}\,dQ^{n,\lambda}. \tag{4.36}$$

Denote

$$G_n(\lambda, S_n) = -\ln E(\mathcal{E}_n^{-1}(\lambda)|S_n).$$

Our main assumption is that there exists a function $G(\lambda, y)$ such that

$$\lim_n n^{-1}G_n(\lambda, ny) = G(\lambda, y). \tag{4.37}$$

Evidently, $G_n(\lambda, y)$ and $G(\lambda)$ are convex in λ, and under Cramér's condition, $G_n(\lambda, y)$ is a smooth in λ function. We denote

$$G_n'(\lambda, y) = \frac{\partial G_n}{\partial \lambda}(\lambda, y).$$

Define

$$I(x) = \sup_{\lambda \in R}\,[\lambda x - G(\lambda, x)],$$

and assume that sup is attained at some point λ^0 for given arbitrary x and so

$$G_n'(\lambda, x) = x.$$

We need to impose some regularity conditions on $G_n(\lambda, y)$ and $G(\lambda, y)$:

(i) (continuity in y) for $y', y'' \in R$ such that $|y' - y''| \le \delta$, $\delta > 0$,

$$\lim_{\delta \to 0} \lim_n |n^{-1}G_n(\lambda, ny') - n^{-1}G_n(\lambda, ny'')| = 0;$$

(ii) (compare with the differentiability of free energy function in Theorem 8.1, Chapter 8)

$$\lim_n \inf_{y > (x+\delta)} \inf_{0<\theta\le 1} [y - n^{-1}G'(\lambda^0 + \theta\mu, ny)] = r(\delta, \lambda^0) > 0,$$

$$\lim_n \inf_{y < (x-\delta)} \inf_{0<\theta\le 1} [y - n^{-1}G'(\lambda^0 + \theta\mu, ny)] = r(\delta, \lambda^0) > 0;$$

(iii) (compare with the item (iv) of Lemma 1.1 in Section 1.1)

$$\lim_{c\to\infty} \lim_n \sup_{\lambda \ge 0} \inf_{y \ge c} [\lambda y - G_n(\lambda, ny)] = \infty,$$

and for $\lambda \le 0$

$$\lim_{c\to\infty} \lim_n \sup_{\lambda \le 0} \inf_{y \le -c} [\lambda y - G_n(\lambda, ny)] = \infty.$$

We now state the main results.

Theorem 4.4. Let conditions (4.37) and (i) hold. Then

$$\limsup_{\delta\to 0}\ \limsup_{n} n^{-1} \ln \mathsf{P}\left(\left|\frac{S_n}{n} - x\right| \le \delta\right) \le -I(x).$$

If, in addition, (iii) holds, then the upper bound of the large deviation principle is valid.

Theorem 4.5. Let conditions (4.37), (i) and (ii) hold. Then

$$\liminf_{\delta\to 0}\ \liminf_{n} n^{-1} \ln \mathsf{P}\left(\left|\frac{S_n}{n} - x\right| \le \delta\right) \ge -I(x).$$

Before getting down to the proof of the main results, it will be useful to consider a simple example. Evidently, the example below could be treated by different methods. Nevertheless, it gives some insight into our approach.

Example. Let $X = (X_k)$, $k \ge 0$, be a sequence, governed by the stochastic difference equation

$$X_k = aX_{k-1} + \xi_k,$$

where $|a| < 1$ and (ξ_k), $k \ge 1$, is a sequence of i.i.d. random variables with a distribution μ and $\mathsf{E}\xi_1 = 0$ such that, for any $\lambda \in R$,

$$\mathsf{E}e^{\lambda\xi_1} < \infty.$$

Denote

$$G_\mu = \ln \mathsf{E}e^{\lambda\xi_1}.$$

We assume that X is adapted to a filtration $\mathbf{F} = (F_k)$, $k \ge 0$. In this case

$$\ln \mathcal{E}_n(\lambda) = \sum_{k=1}^{n} \ln \mathsf{E}\,(e^{\lambda X_k} | F_{k-1}) = \lambda a \sum_{k=1}^{n} X_{k-1} + nG_\mu(\lambda),$$

and

$$\begin{aligned} n^{-1}G_n(\lambda, S_n) &= -n^{-1} \ln \mathsf{E}\,(e^{-\ln \mathcal{E}_n(\lambda)} | S_n) \\ &= G_\mu(\lambda) - n^{-1} \ln \mathsf{E}\left(\exp\left(-\lambda a \sum_{k=1}^{n} X_{k-1}\right) | S_n\right) \\ &= G_\mu(\lambda) + n^{-1}\lambda a \sum_{k=1}^{n} X_k - n^{-1} \ln \mathsf{E}\,(\exp(\lambda a(X_n - X_0)) | S_n). \end{aligned}$$

If for any $y \in R$

$$\lim_{n} n^{-1} \ln \mathsf{E}\,(\exp(\lambda a(X_n - X_0)) | yn) = 0, \tag{4.38}$$

then

$$\lim_{n} n^{-1}G_n(\lambda, yn) = \lambda a y + G_\mu(\lambda).$$

Therefore

$$I(x) = \sup_{\lambda\in R} [\lambda x - \lambda a x - G_\mu(\lambda)] = L((1-a)x), \tag{4.39}$$

where

$$L(x) = \sup_{\lambda \in R} [\lambda y - G_\mu(\lambda)]. \tag{4.40}$$

We assume that sup in (4.40) is attained at some point λ^0 for given arbitrary x and so

$$G'_\mu(\lambda^0) = (1-a)x. \tag{4.41}$$

Let check conditions (i)–(iii). Note that

$$\begin{aligned} |n^{-1}G_n(\lambda, ny') - n^{-1}G_n(\lambda, ny'')| \\ \leq \lambda a|y'-y''| + |n^{-1}\ln \mathsf{E}\,(\exp(\lambda a(X_n - X_0))|y'n) \\ -n^{-1}\ln \mathsf{E}\,(\exp(\lambda a(X_n - X_0))|y''n)|, \end{aligned}$$

and if (4.38) holds then (i) is valid.

We next verify (ii). To this end, we note that

$$\begin{aligned} n^{-1}G'_n(\lambda^0+\theta\mu, ny) &= G'_\mu(\lambda^0+\theta\mu) + ay \\ &\quad -n^{-1}\frac{\mathsf{E}\,(a(X_n - X_0)\, e^{a(\lambda^0+\theta\mu)(X_n - X_0)}|ny)}{\mathsf{E}\,(e^{a(\lambda^0+\theta\mu)(X_n - X_0)}|ny)}. \end{aligned}$$

If, for every λ and y,

$$\lim_n n^{-1}\frac{\mathsf{E}\,(a(X_n - X_0)\, e^{a(\lambda)(X_n - X_0)}|ny)}{\mathsf{E}\,(e^{a(\lambda)(X_n - X_0)}|ny)} = 0,$$

then (see (4.41))

$$\begin{aligned} &\lim_n \inf_{y>(x+\delta)} \inf_{0<\theta\leq 1} [y - n^{-1}G'_n(\lambda^0+\theta\mu, ny)] \\ &= \inf_{y>(x+\delta)} \inf_{0<\theta\leq 1} [(1-a)y - G'_\mu(\lambda^0+\theta\mu)] > 0. \end{aligned}$$

Finally, turning to the condition (iii), we note that

$$\lambda y - n^{-1}G_n(\lambda, ny) = (1-a)y - G_\mu(\lambda) + n^{-1}\ln \mathsf{E}\,(\exp(\lambda a(X_n - X_0))|yn).$$

Thus

$$\lim_n \sup_{\lambda\geq 0} \inf_{y\geq c} [\lambda y - n^{-1}G_n(\lambda, ny)] = \sup_{\lambda\geq 0} [\lambda c - G_\mu(\lambda)],$$

and by (iii) and (iv) of Lemma 1.1, Section 1.1 we get

$$\lim_{c\to\infty} \lim_n \sup_{\lambda\geq 0} \inf_{y\geq c} [\lambda y - n^{-1}G_n(\lambda, ny)] = \infty.$$

Upper bound

We will follow the scheme of Section 1.4, i.e., we will prove the local upper bound estimate and then check the exponential tightness of the sequence S.

Recall that $\mathsf{E}\tilde{z}_n(\lambda) = 1$. Thus

$$\begin{aligned} 1 &\geq \mathsf{E}\Big(\mathrm{I}\left(\left|\frac{S_n}{n} - x\right| \leq \delta\right) \exp\left(n\left[\lambda\frac{S_n}{n} - n^{-1}G_n(\lambda, S_n)\right]\right)\Big) \\ &\geq \mathsf{E}\Big(\mathrm{I}\left(\left|\frac{S_n}{n} - x\right| \leq \delta\right) \exp\big(n[\lambda x - G(\lambda, x)] \\ &\quad -|\lambda|\delta - |G(\lambda, x) - n^{-1}G_n(\lambda, nx)| \\ &\quad -|n^{-1}G_n(\lambda, nx) - n^{-1}G_n(\lambda, S_n)|\big)\Big). \end{aligned} \tag{4.42}$$

In virtue of conditions (4.37) and (i), for any $\varepsilon > 0$, one can choose δ_ε such that for a sufficiently large n on the set $(|S_n/n - x| \leq \delta)$ with $\delta \leq \delta_\varepsilon$ the following inequality takes place:

$$|G(\lambda, x) - n^{-1}G_n(\lambda, nx)| + |n^{-1}G_n(\lambda, nx) - n^{-1}G_n(\lambda, S_n)| \leq \varepsilon.$$

The last inequality and (4.42) imply that for $\delta \leq \delta_\varepsilon$ and for sufficiently large n

$$n^{-1}\ln \mathsf{P}\left(\left|\frac{S_n}{n} - x\right| \leq \delta\right) \leq -[\lambda x - G(\lambda, x)] + |\lambda|\delta + \varepsilon,$$

and since $\varepsilon > 0$ is arbitrary, we get the local upper bound

$$\limsup_{\delta\to 0}\ \limsup_{n}\ n^{-1}\ln \mathsf{P}\left(\left|\frac{S_n}{n} - x\right| \leq \delta\right) \leq -\sup_{\lambda\in R}\,[\lambda x - G(\lambda, x)].$$

We next note that to check the exponential tightness in the finite dimension case it is sufficient to show that

$$\lim_{c\to\infty}\ \lim_{n}\ n^{-1}\mathsf{P}(|S_n| \geq nc) = -\infty.$$

To this end, pick $\lambda \geq 0$ and observe that

$$\begin{aligned} 1 &\geq \mathsf{E}\,\mathrm{I}(S_n \geq nc)\exp\left(n\left[\lambda\frac{S_n}{n} - n^{-1}G_n(\lambda, S_n)\right]\right) \\ &\geq \mathsf{E}\,\mathrm{I}(S_n \geq nc)\exp\left(n\inf_{y\geq c}[\lambda y - n^{-1}G_n(\lambda, ny)]\right). \end{aligned}$$

This inequality implies that

$$n^{-1}\ln \mathsf{P}(S_n \geq nc) \leq -\sup_{\lambda\geq 0}\ \inf_{y\geq c}\,[\lambda y - n^{-1}G_n(\lambda, ny)],$$

and so by condition (iii)

$$\lim_{c\to\infty}\ \lim_{n}\ n^{-1}\mathsf{P}(S_n \geq nc) = -\infty.$$

The set $(S_n \leq -nc)$ is treated similarly. Theorem 4.4 is proved.

Lower bound
We start with the following ergodic lemma. Pick $\lambda^0 = \arg\sup_{\lambda\in R} [\lambda x - G(\lambda, x)]$.

Lemma 4.4. Let condition (ii) hold. Then

$$\lim_n Q^{n,\lambda^0}\left(\left|\frac{S_n}{n} - x\right| > \delta\right) = 0.$$

Proof. It is sufficient to consider $Q^{n,\lambda^0}(S_n/n > x + \delta)$, the set $(S_n/n < x - \delta)$ is treated similarly. Define for some $\mu > 0$

$$\begin{aligned} \tilde{z}_n(\lambda^0, \mu) &= e^{\mu S_n} \frac{\mathsf{E}(\mathcal{E}_n^{-1}(\lambda^0 + \mu)|S_n)}{\mathsf{E}(\mathcal{E}_n^{-1}(\lambda^0)|S_n)} \\ &= \exp\left(n\left[\mu\frac{S_n}{n} - n^{-1}G_n(\lambda^0 + \mu, S_n) + n^{-1}G_n(\lambda^0, S_n)\right]\right), \end{aligned}$$

and note that

$$\mathsf{E}^{n,\lambda^0}\tilde{z}_n(\lambda^0, \mu) = 1. \tag{4.43}$$

Taking into account that $G_n(\lambda, S_n)$ is a smooth in λ function, for some θ, $0 < \theta \leq 1$ from (4.43) we obtain

$$\begin{aligned} 1 &\geq \mathsf{E}^{n,\lambda^0}\mathrm{I}\left(\frac{S_n}{n} > x + \delta\right)\exp\left(n\left[\mu\frac{S_n}{n} - n^{-1}G_n(\lambda^0 + \mu, S_n) + n^{-1}G_n(\lambda^0, S_n)\right]\right) \\ &= \mathsf{E}^{n,\lambda^0}\mathrm{I}\left(\frac{S_n}{n} > x + \delta\right)\exp\left(n\left[\mu\frac{S_n}{n} - \mu n^{-1}G_n'(\lambda^0 + \theta\mu, S_n)\right]\right) \\ &\geq \mathsf{E}^{n,\lambda^0}\mathrm{I}\left(\frac{S_n}{n} > x + \delta\right)\exp\left(n\mu \inf_{y\geq(x+\delta)}\inf_{0<\theta\leq 1}[y - n^{-1}G_n'(\lambda^0 + \theta\mu, ny)]\right). \end{aligned}$$

This inequality implies

$$n^{-1}\ln Q^{n,\lambda^0}\left(\frac{S_n}{n} > x + \delta\right) \leq -\mu \inf_{y\geq(x+\delta)}\inf_{0<\theta\leq 1}[y - n^{-1}G_n'(\lambda^0 + \theta\mu, ny)],$$

and so, by condition (ii) we get the statement of Lemma 4.4.

Turning to the proof of lower bound, from (4.36) we obtain

$$\begin{aligned} \mathsf{P}\left(\left|\frac{S_n}{n} - x\right| \leq \delta\right) &= \mathsf{E}^{n,\lambda^0}\mathrm{I}\left(\left|\frac{S_n}{n} - x\right| \leq \delta\right)\exp\left(-n\left[\lambda^0\frac{S_n}{n} - n^{-1}G_n(\lambda^0, S_n)\right]\right) \\ &\geq \mathsf{E}^{n,\lambda^0}\mathrm{I}\left(\left|\frac{S_n}{n} - x\right| \leq \delta\right)\exp(-n[\lambda^0 x - G(\lambda^0, x) + |\lambda^0|\delta \\ &\quad + |G(\lambda^0, x) - n^{-1}G_n(\lambda^0, nx)| + |n^{-1}G_n(\lambda^0, S_n) \\ &\quad - n^{-1}G_n(\lambda^0, nx)|]). \end{aligned} \tag{4.44}$$

In virtue of conditions (4.37) and (i), for any $\varepsilon > 0$, one can choose δ_ε such that for a sufficiently large n on the set $(|S_n/n - x| \leq \delta)$ with $\delta \leq \delta_\varepsilon$ the following inequality takes place:

$$|G(\lambda^0, x) - n^{-1}G_n(\lambda^0, nx)| + |n^{-1}G_n(\lambda^0, nx) - n^{-1}G_n(\lambda^0, S_n)| \leq \varepsilon.$$

The last inequality and (4.44) imply that for $\delta \le \delta_\varepsilon$ and sufficiently large n

$$n^{-1} \ln \mathsf{P}\left(\left|\frac{S_n}{n} - x\right| \le \delta\right) \ge -[\lambda^0 x - G(\lambda^0, x)] - \lambda^0 \delta - \varepsilon,$$

and since $\varepsilon > 0$ is arbitrary, we get the lower bound

$$\liminf_{\delta \to 0} \liminf_{n} n^{-1} \ln \mathsf{P}\left(\left|\frac{S_n}{n} - x\right| \le \delta\right) \ge -I(x).$$

Comments

This chapter is the result of one of the authors' attempt (O.G), partly to extend Cramér's theory to a sequence of semimartingales under conditions, expressed in terms of the stochastic exponent. These results could be extended to an infinite dimension case. Unfortunately, we do not know a good example which could make us to do that.

Chapter 5

Mixing for Markov processes

5.1. DEFINITIONS

The chapter contains some auxiliary but having independent value results concerning some bounds of a mixing rate for a class of Markov processes which we make use of next to prove the averaging principle and the normal deviations for the processes with averaging (see Chapters 6 and 7).

Let us consider the Markov process X_t on some complete probability space (Ω, F, P) with values in R^d. Time is discrete, $t \in \mathbf{Z}^1_+$. We propose the method which allows us to obtain the estimates for the following mixing coefficients:

Rosenblatt's coefficient of strong mixing

$$\alpha(t) = \sup_{s\geq 0} \sup_{A\in F^X_{\leq s}} \sup_{B\in F^X_{\geq s+t}} |\mathsf{P}(AB) - \mathsf{P}(A)\,\mathsf{P}(B)|;$$

Kolmogorov's complete regularity coefficient

$$\beta(t) = \sup_{s\geq 0} \operatorname*{var}_{B\in F^X_{\geq s+t}} (\mathsf{P}(B|F^X_{\leq s}) - \mathsf{P}(B));$$

Ibragimov's uniformly strong mixing coefficient

$$\varphi(t) = 2 \sup_{s\geq 0} \operatorname*{ess}_{\mathsf{P}} \sup_{B\in F^X_{\geq s+t}} |\mathsf{P}(B|F^X_{\leq s}) - \mathsf{P}(B)|;$$

ψ-coefficient

$$\psi(t) = 2 \sup_{s\geq 0} \operatorname*{var}_{B\in F^X_{\geq s+t}} |\mathsf{P}(B|F^X_{\leq s})\,\mathsf{P}(B)^{-1} - 1|;$$

χ-coefficient

$$\chi(t) = 2 \sup_{s\geq 0} \operatorname*{ess}_{\mathsf{P}} \sup_{B\in F^X_{\geq s+t}} |\mathsf{P}(B|F^X_{\leq s})\,\mathsf{P}(B)^{-1} - 1|;$$

the variance distance

$$v^{x,y}(t) = \operatorname{var}\,(\mathsf{P}^x_t - \mathsf{P}^y_t)$$

(here P^x_t is a distribution of X_t under condition $X_0 = \mathbf{x}$). The following inequalities hold:

$$\alpha(t) \leq \beta(t) \leq \varphi(t) \leq \chi(t), \qquad \alpha(t) \leq \beta(t) \leq \psi(t) \leq \chi(t).$$

Let us state the stability conditions which will be used in this chapter. Further, a is a positive number, $K \in \mathcal{B}(R^d)$, $\tau = \inf\,(t \geq 0\colon X_t \in K)$, ϑ is a shift operator of our Markov process.

(A_K^a) $\quad \sup_{s\geq 0} \mathsf{E}_{s,x} \sup (a\vartheta_s T) \leq h(x),$

where h is some non-negative function bounded on $K*$;

(B_K^a) $\quad \sup_{s,t\geq 0} \mathsf{E}_{s,x} h(X_t) \leq C_K h(x);$

$$(D_K) \qquad q_K = \inf_t \inf_{x,x'\in K} \int_K \left(1 \wedge \frac{\mathsf{P}(t,x',t+1,dy)}{\mathsf{P}(t,x,t+1,dy)}\right) \mathsf{P}(t,x,t+1,dy) > 0,$$

where $d\mathsf{P}'/d\mathsf{P}$ means the density of the absolute continuous component of one measure with respect to another;

$(D_{R^d}) = (D_K)$ as $K = R^d$;

(D_b) measures $\mathsf{P}(t,x,t+1,dy)$ and $\mathsf{P}(t,x',t+1,dy)$ are equivalent for any $x, x' \in R^d$ and $t \in \mathbf{Z}^1_+$ and

$$0 < \nu^{-1} \leq \inf_t \inf_{x,x'} \frac{\mathsf{P}(t,x',t+1,dy)}{\mathsf{P}(t,x,t+1,dy)} \leq \sup_t \sup_{x,x'} \frac{\mathsf{P}(t,x',t+1,dy)}{\mathsf{P}(t,x,t+1,dy)} \leq \nu < \infty;$$

(D_f) measures $\mathsf{P}(t,x,t+1,dy)$ and $\mathsf{P}(t,x',t+1,dy)$ are equivalent for any $x, x' \in R^d$ and $t \in \mathbf{Z}^1_+$ and inequalities

$$\sup_t \frac{\mathsf{P}(t,x',t+1,dy)}{\mathsf{P}(t,x,t+1,dy)} \leq f(x,x') < \infty,$$

and

$$\sup_{s,t} \mathsf{E}_{s,x} f(X_t, X'_t) < \infty$$

hold true for some function $f\colon R^d \times R^d \to R^1_+$, where X_t and X' are independent versions of initial Markov process and E means the expectation on the direct product of two correspondent probability spaces.

5.2. MAIN RESULTS

Theorem 5.1. Let conditions $(A_K^a) + (B_K^a) + (D_K)$ with $a > 0$, $K \subset R^d$, be satisfied. Then there exist such $C, \lambda > 0$, that for any x, t

$$\alpha(t) \leq \beta(t) \leq C \exp(-\lambda t)\, h^2(x). \tag{5.1}$$

Theorem 5.2. Let condition (D_{R^d}) be satisfied. Then there exist such $C, \lambda > 0$ that for any x, t

$$\varphi(t) \leq C \exp(-\lambda t). \tag{5.2}$$

Theorem 5.3. Let condition (D_b) be satisfied. Then there exist such $C, \lambda > 0$ that for any x, t

$$\psi(t) \leq \chi(t) \leq C \exp(-\lambda t). \tag{5.3}$$

Theorem 5.4. Let condition (D_f) be satisfied. Then there exist such $C, \lambda > 0$ that for any x, t

$$\psi(t) \le C \exp(-\lambda t). \tag{5.4}$$

The next theorem deals with convergence in variance.

Theorem 5.5. (1) Let $(A_K^a) + (B_K^a) + (D_K)$ with $a > 0$, $K \subset R^d$, be satisfied. Then there exist such $C, \lambda > 0$, that for any x, y, t

$$v^{x,y}(t) \le C \exp(-\lambda t)\, h(x)\, h(y). \tag{5.5}$$

(2) If (D_{R^d}) holds then for some $C, \lambda > 0$ and any x, x', t

$$v^{x,y}(t) \le C \exp(-\lambda t). \tag{5.6}$$

In particular if the Markov process X_t is homogeneous then under conditions of Theorem 5.5 (1) or (2) X_t has a unique invariant measure P_∞, and if $\int h(y)\, \mathsf{P}_\infty(dy) < \infty$, then (5.5) implies inequality

$$\operatorname{var}(\mathsf{P}_t^x - \mathsf{P}_\infty) \le C \exp(-\lambda t)\, h(x) \tag{5.7}$$

with some $C > 0$. Under conditions of Theorem 5.5 (2) we obtain

$$\sup_x \operatorname{var}(\mathsf{P}_t^x - \mathsf{P}_\infty) \le C \exp(-\lambda t). \tag{5.8}$$

(5.8) is a special case of results from Doob (1953), Chapters 5 and 6 on a convergence rate to invariant measure. (However, previous estimates are not.)

Proofs of all theorems are based on a unique scheme with the use of a coupling method. Thus, we prove in detail two theorems only, Theorem 5.1 and Theorem 5.4. Theorem 5.2 is the important special case of Theorem 5.1.

Our method allows us to obtain polynomial estimates for mixing coefficients as well. Here we state only one result of such type.

Let us introduce two more stability conditions for X_t.

$(A_{K,n})$ $\quad \sup_{s \ge 0} \mathsf{E}_{s,x}(\vartheta_s T)^n \le h(x);$

$(B_{K,n})$ $\quad \sup_{s,t} \mathsf{E}_{s,x} h(x_t) \le C h(x),$

where $h(x)_K \equiv 1$.

Theorem 5.6. Let $(A_{K,n}) + (B_{K,n})$ with some $K \in \mathcal{B}^d$, $n \ge 0$, be satisfied. Then

$$\alpha(t) \le \beta(t) \le C(1+t)^{-n} (\ln t)^{n-1} h^2(x).$$

5.3. PRELIMINARY RESULTS

Fix $s \ge 0$, and let $(\xi_t,\ t \ge s)$, $(\eta_t,\ t \ge 0)$ be independent copies of our Markov process X_t with the same transition probabilities and initial values $\xi_s = X_s$ and

$\eta_0 = X_0$. We will built a new Markov process $(\zeta_t,\ t \geq 0)$ on the direct product of correspondent probability spaces such that ζ_t coincides with η_t until some Markov moment (the moment of coupling) and coincides with ξ_t after that moment. That construction allows us to prove Theorems 5.1–5.6.

Denote $T_1 = \inf(t \geq s\colon \xi_t \in K,\ \eta_t \in K)$, $T_{i+1} = \inf(t \geq T_i + 1\colon \xi_t \in K,\ \eta_t \in K)$, $i \geq 1$; here K is the set from (A_K^a), (B_K^a), or (D_K). Of course, in cases (D_{R^d}) and (D_b) we have $T_1 = s$, $T_2 = s + 1, \ldots$.

Lemma 5.1. If $(A_K^a) + (B_K^a) + (D_K)$ hold with some $a > 0$, $K \subset R^d$, then for any $\delta > 0$ small enough, any $X_0, i = 1,2,\ldots$, and some $C > 0$

$$\mathsf{E}(\exp(\delta(T_1 - s))|\xi_s) \leq Ch(X_0)\,h(\xi_s), \tag{5.9}$$

$$\mathsf{E}\exp(\delta(T_1 - s)) \leq Ch^2(X_0), \tag{5.10}$$

$$\mathsf{E}(\exp(\delta(T_{i+1} - T_i))|\xi_s) \leq C. \tag{5.11}$$

The value $C - 1 > 0$ here may be chosen arbitrary small if $\delta > 0$ is small enough.

Proof. The last assertion about $C \cong 1$ follows from other ones and from the Hölder inequality: as $h \geq 1$, then

$$\mathsf{E}\exp((\delta/n)(T_1 - s)) \leq C^{1/n}h^{2/n}(X_0) \leq C^{1/n}h^2(X_0).$$

(5.10) follows from (5.9) with another C and from (B_K^a). (5.11) follows from (5.10), as $h_K \leq \text{const}$. So let us prove (5.9). Since (D_K) holds we have

$$q_1 = \inf_{y \in K} \mathsf{P}_y(X_1 \in K) > 0,$$

and by virtue of the Markovian property this implies

$$q_n = \inf_{y \in K} \mathsf{P}_y(X_n \in K) \geq q_1^n > 0.$$

By virtue of (A_K^a) and (B_K^a) moments $\tau^\xi = \tau_0^\xi = \inf(t \geq s + 1\colon \xi_t \in K)$ and $\tau^\eta = \tau_0^\eta = \inf(t \geq \tau_0^\xi\colon \eta_t \in K)$ are finite with probability 1. Since two processes are independent and by virtue of (B_K^a) we obtain

$$\mathsf{E}_{\tau_0^\xi, \xi_{\tau_0^\xi}} h_K(\xi_{\tau_1^\eta}) \leq C.$$

Further, let $\xi_{\tau_1^\eta} \notin K$. Denote $\tau_1^\xi = \inf(t \geq \tau_1^\eta\colon \xi_t \in K)$. By virtue of (A_K^a) and the Chebyshov inequality we obtain

$$\begin{aligned} \mathsf{P}(\tau_1^\xi - \tau_1^\eta \geq n|\tau_1^\xi, \xi_{\tau_1^\eta}) &\leq \exp(-\alpha n)\,\mathsf{E}(\exp(\alpha(\tau_1^\xi - \tau_1^\eta))|\tau_1^\xi, \xi_{\tau_1^\eta}) \\ &\leq \exp(-\alpha n)\,h(\xi_{\tau_1^\eta}). \end{aligned}$$

Hence,

$$\mathsf{P}(\tau_1^\xi - \tau_1^\eta \geq n) \leq \exp(-\alpha n)\,\mathsf{E}h(\xi_{\tau_1^\eta}) \leq C\exp(-\alpha n).$$

Let us denote $\tau_{i+1}^{\eta} = \inf(t \geq \tau_i^{\xi}: \eta_t \in K)$, $\tau_{i+1}^{\xi} = \inf(t \geq \tau_{i+1}^{\eta}: \xi_t \in K)$. In the same way we obtain

$$\begin{aligned} \mathsf{P}(\tau_{i+1}^{\eta} - \tau_i^{\xi} \geq n | \tau_i^{\xi}, \eta_{\tau_i^{\xi}}) &\leq C \exp(-\alpha n), \\ \mathsf{P}(\tau_i^{\xi} - \tau_i^{\eta} \geq n | \tau_i^{\eta}, \xi_{\tau_i^{\eta}}) &\leq C \exp(-\alpha n), \qquad i \geq 1. \end{aligned}$$

Moreover,

$$\mathsf{P}(\tau_1^{\eta} - \tau_0^{\xi} \geq n | s, \xi_s) \leq C \exp(-\alpha n) \,\mathsf{E}\,(h(\eta_{\tau_s^{\xi}}) | s, \xi_s) \leq C \exp(-\alpha n)\, h(X_0),$$

and

$$\mathsf{P}(\tau_0^{\xi} - s \geq n | s, \xi_s) \leq C \exp(-\alpha n)\, h(\xi_s).$$

Let us choose such n that $C \exp(-\alpha n) \leq h(\xi_s)$. For any n, and hence for fixed one for any k the estimates hold:

$$\mathsf{P}(\xi_i \in K, \tau_k^{\xi} \leq i \leq \tau_k^{\xi} + n) \geq q^n > 0,$$

$$\mathsf{P}(\eta_i \in K, \tau_k^{\eta} \leq i \leq \tau_k^{\eta} + n) \geq q^n > 0.$$

Thus, for any $i \geq 0$

$$\begin{aligned} \mathsf{P}(\tau_{i+1}^{\xi} = \tau_{i+1}^{\eta}) &\geq q^n/2, \\ \mathsf{P}(\tau_{i+2}^{\eta} = \tau_{i+1}^{\xi}) &\geq q^n/2. \end{aligned}$$

Let us denote $\Delta_0 = \tau_0^{\xi} - s$, $\Delta_1 = \tau_1^{\eta} - \tau_0^{\xi}$, $\Delta_2 = \tau_1^{\xi} - \tau_1^{\eta}$, $\Delta_3 = \tau_2^{\eta} - \tau_1^{\xi}, \ldots$ For $\delta \leq \alpha$ we have

$$\begin{aligned} \mathsf{E}\,(\exp(\delta\Delta_0) | s, \xi_s) &\leq h(\xi_s), \\ \mathsf{E}\,(\exp(\delta\Delta_1) | F_{s+\Delta_0}^{\xi,\eta}) &\leq C h(X_0), \\ \mathsf{E}\,(\exp(\delta\Delta_1) | F_{s+\Delta_0+\Delta_1+\ldots+\Delta_{i-1}}^{\xi,\eta}) &\leq C, \qquad i \geq 2, \end{aligned}$$

and moreover,

$$\mathsf{P}(\Delta_i = 0 | F_{s+\Delta_0+\Delta_1+\ldots+\Delta_{i-1}}^{\xi,\eta}) \geq q^n/2 > 0, \qquad i \geq 1.$$

By virtue of Borel–Cantelli lemma on the set of P-measure 1

$$T_i - s \leq \sum_{i=0}^{\infty} \Delta_i = \sum_{i=1}^{\inf(k \geq 0:\ \Delta_k = 0)} \Delta_i.$$

Hence,

$$\begin{aligned} \mathsf{E}\,(\exp(\delta(T_1 - s)) | s, \xi_s) &\leq \sum_{k=1}^{\infty} (1 - q^n/2)^{k-1} \mathsf{E}\left(\exp\left(\delta \sum_{i=0}^{k-1} \Delta_i\right) | s, \xi_s\right) \\ &\leq \sum_{k=1}^{\infty} (1 - q^n/2)^{k-1} C^{k-1} h(X_0)\, h(\xi_s). \end{aligned}$$

Note that for $\alpha > 0$ small enough the constant C in (B_K^{α}) may be chosen arbitrary near 1. Indeed, by virtue of the Hölder inequality from $\mathsf{E} \exp(\alpha\tau) \leq h(x) (\geq 1)$ and

$\mathsf{E}\,h(X_i) \le Ch(x)$ follows $\mathsf{E}\exp((\alpha/m)\tau) \le h(x)^{1/m}$ and $\mathsf{E}\,h(X_i)^{1/m} \le C^{1/m}h(x)^{1/m}$ for any natural m; when changing α to α/m one must also change h to $h^{1/m}$. Thus, for $\alpha > 0$ small enough the right-hand side in (2.4) is finite. This proves (5.9) together with all other assertions. Lemma 5.1 is proved.

Let us state one more preliminary result which is really the base of the coupling method (that is, the method of a unique probability space). One may read about the coupling method in Nummelin (1984) (see Remarks below). The result—Lemma 5.2—may be considered as a folk-lore.

Let $(\Omega_1, F_1, \mathsf{P}_1)$ and $(\Omega_2, F_2, \mathsf{P}_2)$ be two independent probability spaces, and ξ_1: $(\Omega_1, F_1) \to (R^1, \mathcal{B}^1)$, ξ_2: $(\Omega_2, F_2) \to (R^1, \mathcal{B}^1)$ be two random values on $(\Omega_1, F_1, \mathsf{P}_1)$ and $(\Omega_2, F_2, \mathsf{P}_2)$ with distributions $\mathcal{L}(\xi_1)$ and $\mathcal{L}(\xi_2)$. Let $\mathcal{L}(\xi_i) \ll \Lambda$ for some measure Λ on $(R^1, \mathcal{B}^1)$, and

$$q = \int \min(\mathrm{d}\mathcal{L}(\xi_1)/\,\mathrm{d}\Lambda,\ \mathrm{d}\mathcal{L}(\xi_2)/\,\mathrm{d}\Lambda)\,\mathrm{d}\Lambda.$$

Lemma 5.2. If $q > 0$ then there exists such an extension $(\Omega_3, F_3, \mathsf{P}_3)$ of probability space $(\Omega_2, F_2, \mathsf{P}_2)$ and such a random value ξ_3: $(\Omega_3, F_3) \to (R^1, \mathcal{B}^1)$, that $\mathcal{L}(\xi_3) = \mathcal{L}(\xi_1)$, and

$$\mathsf{P}_3(\xi_3 = \xi_2) \ge q > 0.$$

(Here ξ_2 is considered as the random value on $(\Omega_3, F_3, \mathsf{P}_3)$.)

Proof. Let us denote $p_i(x) = \mathrm{d}\mathcal{L}(\xi_i)/\,\mathrm{d}\Lambda(x)$, $i = 1, 2$. Let ζ be a random value with uniform distribution on [0, 1] and independent from ξ_1 and ξ_2. Let us define

$$\xi_3 = \xi_2 \mathrm{I}\left(\frac{p_1(\xi_2)}{p_1 \vee p_2(\xi_2)} \ge \zeta\right) + \eta\,\mathrm{I}\left(\frac{p_1(\xi_2)}{p_1 \vee p_2(\xi_2)} < \zeta\right),$$

where ζ is one more independent random value with a density

$$p_\eta(x) = (p_1 - p_1 \wedge p_2)(x)/\int (p_1 - p_1 \wedge p_2)\,\mathrm{d}\Lambda.$$

w.r.t. Λ. Then $\mathcal{L}(\xi_3) = \mathcal{L}(\xi_1)$, and

$$\mathsf{P}_3(\xi_3 = \xi_2) \ge \mathsf{P}_3\left(\frac{p_1(\xi_2)}{p_1 \vee p_2(\xi_2)} \ge \zeta\right) = \int p_1 \wedge p_2\,\mathrm{d}\Lambda = q > 0.$$

Lemma 5.2 is proved.

Lemma 5.3. If $(D_K) + (A_{K,n}) + (B_{K,n})$ hold with some $K \in \mathcal{B}^d$, $n \ge 0$, then for any $X_0, i = 1, 2, \ldots$ and some $C > 0$ the following estimates hold

$$\mathsf{E}\,((T_1 - s)^n | s, \xi_s) \le Ch(X_0)\,h(\xi_s), \tag{5.12}$$

$$\mathsf{E}\,(T_1 - s)^n \le Ch^2(X_0), \tag{5.13}$$

$$\mathsf{E}\,((T_{i+1} - T_i)^n | F^{\xi,\eta}_{T_i}) \le C. \tag{5.14}$$

Proof. We have

$$\begin{aligned}
\mathsf{E}((T_1-s)^k|s,\xi_s) &\le \mathsf{E}\left(\left(\sum_{i=0}^{\infty}\Delta_i\right)^k|s,\xi_s\right)\\
&\le \sum_{r=1}^{\infty}(1-q^n/2)^{r-1}\mathsf{E}\left(\left(\sum_{i=0}^{r-1}\Delta_i\right)^k|s,\xi_s\right)\\
&\le \sum_{r=1}^{\infty}(1-q^n/2)^{r-1}r^k\mathsf{E}\left(\left(\sum_{i=0}^{r-1}\Delta_i^k\right)|s,\xi_s\right)\\
&\le \sum_{r=1}^{\infty}(1-q^n/2)^{r-1}r^kC^r h(\xi_s)\,h(x_0),
\end{aligned}$$

which implies (5.12) and (5.13). (5.14) is proved in the same way. Lemma 5.3 is proved.

5.4. PROOFS OF THEOREMS 5.1-5.6

Let $\mathcal{B}^d_{[a,b]}$ be a Borel σ-field in $C[a,b;R^d]$. Let us denote by ξ_a^b the path of the process ξ_t when $a \le t \le b$. Let $\Gamma \in \mathcal{B}^d_{[s+t,\infty)}$ be such that $\mathsf{P}(\eta^{\infty}_{s+t} \in \Gamma) > 0$. By virtue of the Markovian property and independence of ξ and η we have

$$\mathsf{P}(\xi^{\infty}_{s+t} \in \Gamma|s,\xi_s) = \mathsf{P}(\xi^{\infty}_{s+t} \in \Gamma|F^{\xi}_s),$$

$$\mathsf{P}(\eta^{\infty}_{s+t} \in \Gamma) = \mathsf{E}(\mathrm{I}(\eta^{\infty}_{s+t} \in \Gamma)|F^{\xi}_s),$$

thus, for any n

$$\begin{aligned}
\left|\frac{\mathsf{P}(\xi^{\infty}_{s+t} \in \Gamma|s,\xi_s)}{\mathsf{P}(\eta^{\infty}_{s+t} \in \Gamma)}-1\right| &= \mathsf{P}(\eta^{\infty}_{s+t} \in \Gamma)^{-1}|\mathsf{E}\{\mathrm{I}(\xi^{\infty}_{s+t} \in \Gamma)-\mathrm{I}(\eta^{\infty}_{s+t} \in \Gamma)|s,\xi_s\}|\\
&\le N\mathsf{P}(T_n > t+s|F^{\xi}_s) + \mathsf{P}(\eta^{\infty}_{s+t} \in \Gamma)^{-1}\\
&\quad \times|\mathsf{E}\{\mathrm{I}(T_n \le t+s)(\mathrm{I}(\xi^{\infty}_{s+t} \in \Gamma)-\mathrm{I}(\eta^{\infty}_{s+t} \in \Gamma))|F^{\xi}_s\}|,
\end{aligned}$$

by virtue of (D_b), where $N = \nu + 1$. Further,

$$\sup_{\Gamma} \mathsf{P}(\eta^{\infty}_{s+t} \in \Gamma)^{-1}|\mathsf{E}\{\mathrm{I}(T_n \le t+s)(\mathrm{I}(\xi^{\infty}_{s+t} \in \Gamma)-\mathrm{I}(\eta^{\infty}_{s+t} \in \Gamma))|F^{\xi}_s\}|$$

$$\le \sup_{\Gamma} \mathsf{P}(\eta^{\infty}_{s+t} \in \Gamma)^{-1}|\mathsf{E}\{\mathrm{I}(T_n \le t+s)(\mathrm{I}(\xi^{\infty}_{T_n} \in \Gamma)-\mathrm{I}(\eta^{\infty}_{T_n} \in \Gamma))|F^{\xi}_s\}|.$$

Let us build some new process ζ_t from ξ_t and η_t with the help of the following construction. The process ζ_s coincides with η_s up to the moment T_1. Now, if ζ_s is already constructed up to the moment T_k, then let us consider its distribution in the moment T_k+1: $\mathsf{P}(T_k,\zeta_{T_k},T_k+1,dy)$. By virtue of Lemma 5.2 and condition (D_b) it is possible to build such a random value $\zeta(T_k+1)$ on some extension of the initial probability space that

$$\mathcal{L}(\zeta(T_k+1)|F^{\xi,\eta}_{T_k}) = \mathcal{L}(\eta(T_k+1)|F^{\xi,\eta}_{T_k})$$

and

$$\mathsf{P}(\zeta(T_k + 1) = \xi_{T_k+1} | F_{T_k}^{\xi,\eta}) \geq q_0 > 0.$$

Let $\zeta_{T_k+1} = \xi_{T_k+1}$ and $\zeta_{T_k+1}^{\infty} = \xi_{T_k+1}^{\infty}$ on the set $(\zeta(T_k + 1) = \xi_{T_k+1})$. Now, let $\zeta_{T_k+1} = \eta_{T_k+1}$, $\zeta_t = \eta_t$ as $t \in [T_k, T_k + 1]$ on the set $(\zeta(T_k + 1) \neq \xi_{T_k+1})$, and repeat all this construction in the moment $T_{k+1} \geq T_k + 1$. On the set $(\zeta(T_k + 1) = \xi_{T_k+1})$ we do not define the path of ζ_t, $t \in (T_k, T_k + 1)$, although it is possible to do with the help of regular conditional distribution. Let us denote $L = \inf\,(T_k + 1\colon \zeta(T_k + 1) = \xi_{T_k+1})$. We have

$$\begin{aligned}
&\sup_{\Gamma} \mathsf{P}(\eta_{s+t}^{\infty} \in \Gamma)^{-1} |\mathsf{E}\{\mathrm{I}(T_n \leq t + s)(\mathrm{I}(\xi_{T_n}^{\infty} \in \Gamma) - \mathrm{I}(\eta_{T_n}^{\infty} \in \Gamma)) | F_s^{\xi}\}| \\
&\quad \leq N\mathsf{P}(L > T_{n-1} | F_s^{\xi}) \\
&\qquad + \sup \mathsf{P}(\eta_{s+t}^{\infty} \in \Gamma) |\mathsf{E}\{\mathrm{I}(T_n \leq t + s; L \leq T_{n-1})(\mathrm{I}(\xi_{T_n}^{\infty} \in \Gamma) - \mathrm{I}(\eta_{T_n}^{\infty} \in \Gamma)) | F_s^{\xi}\}|.
\end{aligned}$$

Since L is the Markovian moment then

$$\begin{aligned}
&\mathsf{E}\{\mathrm{I}(T_n \leq t + s; L \leq T_{n-1})(\mathrm{I}(\xi_{T_n}^{\infty} \in \Gamma) - \mathrm{I}(\eta_{T_n}^{\infty} \in \Gamma)) | F_s^{\xi}\} \\
&\quad = \mathsf{E}\{\mathsf{E}\{\mathrm{I}(T_n \leq t + s; L \leq T_{n-1})(\mathrm{I}(\xi_{T_n}^{\infty} \in \Gamma) - \mathrm{I}(\eta_{T_n}^{\infty} \in \Gamma)) | F_L^{\xi,\eta}\} | F_s^{\xi}\} \\
&\quad = \mathsf{E}\{\mathrm{I}(T_n \leq t + s; L \leq T_{n-1})(\mathsf{P}(\xi_{T_n}^{\infty} \in \Gamma | F_L^{\xi,\eta}) - \mathsf{P}(\eta_{T_n}^{\infty} \in \Gamma | F_L^{\xi,\eta})) | F_s^{\xi}\} \\
&\quad = \mathsf{E}\{\mathrm{I}(T_n \leq t + s; L \leq T_{n-1})(\mathsf{P}(\xi_{T_n}^{\infty} \in \Gamma | F_L^{\xi,\eta}) - \mathsf{P}(\zeta_{T_n}^{\infty} \in \Gamma | F_L^{\xi,\eta})) | F_s^{\xi}\} \\
&\quad = \mathsf{E}\{\mathsf{E}\{\mathrm{I}(T_n \leq t + s; L \leq T_{n-1})(\mathrm{I}(\xi_{T_n}^{\infty} \in \Gamma) - \mathrm{I}(\zeta_{T_n}^{\infty} \in \Gamma)) | F_L^{\xi,\eta}\} | F_s^{\xi}\} = 0.
\end{aligned}$$

Thus,

$$\left| \frac{\mathsf{P}(\xi_{t+s}^{\infty} \in \Gamma | F_s^{\xi})}{\mathsf{P}(\eta_{t+s}^{\infty} \in \Gamma)} - 1 \right| \leq N\{\mathsf{P}(L > T_{n-1} | F_s^{\xi}) + \mathsf{E}(\mathsf{P}(T_n > t + s | F_s^{\xi,\eta}) | F_s^{\xi})\}.$$

Constant n may be chosen here. Let $n = [at + 1]$, then

$$\mathsf{P}(L > T_{n-1} | F_s^{\xi}) \leq (1 - q_0)^{n-1} = (1 - q_0)^{[at]},$$

and

$$\begin{aligned}
\mathsf{P}(T_n > t + s | F_s^{\xi,\eta}) &\leq \exp(-\delta t)\, \mathsf{E}(\exp(\delta(T_n - s)) | F_s^{\xi,\eta}) \\
&\leq \exp(-\delta t)\, C^n h(\xi_s) h(\eta_s) \leq \exp(-\delta t)\, C^{[at+1]} h(\xi_s) h(\eta_s) \\
&\leq \exp(-\delta t + (at + 1) \ln C) h(\xi_s) h(\eta_s).
\end{aligned}$$

Let us choose such $a > 0$ that $a \ln C < \delta/2$. Then

$$\mathsf{P}(T_n > t + s | F_s^{\xi,\eta}) \leq C \exp(-\delta t/2) h(\xi_s) h(h_s).$$

Thus,

$$\begin{aligned}
\mathsf{E} \sup_{\Gamma} \left| \frac{\mathsf{P}(\xi_{s+t}^{\infty} \in \Gamma | s, \xi_s)}{\mathsf{P}(\eta_{s+t}^{\infty} \in \Gamma)} - 1 \right| &\leq N\{(1 - q_0)^{[at]} + C \exp(-\delta t/2)\, \mathsf{E} h(\xi_s) h(\eta_s)\} \\
&\leq N\{(1 - q_0)^{[at]} + C^3 \exp(-\delta t/2) h^2(x)\} \\
&\leq N(C^3 + (1 - q_0)^{-1}) h^2(X_0) \exp(-\lambda t),
\end{aligned}$$

where $\lambda = (\delta/2) - a \ln(1 - q_0)$. Theorem 5.3 is proved.

Proof of Theorem 5.1 follows from the same considerations if we estimate the values $|\mathsf{P}(\xi_{s+t}^{\infty} \in \Gamma | F_s^{\xi}) - \mathsf{P}(\eta_{s+t}^{\infty} \in \Gamma)|$ and take the expectation.

Proof of Theorem 5.4 follows from the similar estimates.

In the proof of Theorem 5.2 we estimate the same values with $T_n = n$ since $\mathsf{P}(T_n > n) = 0$.

Proof of Theorem 5.5 follows by the similar way.

Proof of Theorem 5.6 follows from similar considerations with the help of Lemma 5.3.

5.5. MIXING COEFFICIENTS FOR RECURSIVE PROCEDURES

Let us consider the process X which satisfies the stochastic difference equation

$$X_{n+1} = X_n + a(X_n) + b(X_n)W_n, \qquad n \geq 0.$$

Let the following conditions hold:

A. $\sup (|a(x)| + \|b(x)\|) \leq \nu^{-1}$,

B. $(a(x), x) \leq -\nu|x|, \quad |x| \geq R$,

C. W_n are i.i.d., $\mathsf{E}W_n = 0$, $\mathsf{E} \exp(\varkappa |W_n|) < \infty (\varkappa > 0)$,

D. there exists such $R_1 > R$ and $\nu(R_1) > 0$ that

$$\int_{|z| \leq R} 1 \wedge q_{z_1, z_2, y_1, y_2}(z)\, \mathsf{P}_{z_2, y_2}(\mathrm{d}z) > \nu(R_1)$$

for any $|y_1|, |y_2|, |z_1|, |z_2| \leq R_1$, where $q_{z_1, z_2, y_1, y_2}(z)$ is the density of absolute continuous component of P_{z_1, y_1} w.r.t. P_{z_2, y_2}, and $\mathsf{P}_{z,y}$ is the distribution of the random value $z + b(y)W_n$.

Theorem 5.7. Let conditions A–D hold. Then process X satisfies the conditions of Theorem 5.1 with $h(x) = \exp(\varepsilon|x|)$, $\varepsilon > 0$, and hence

$$\alpha(n) \leq \beta(n) \leq C \exp(-\lambda n) \exp(2\varepsilon|x|).$$

Moreover,

$$\operatorname{var}(\mathsf{P}_n^x - \mathsf{P}_\infty) \leq C \exp(-\lambda n) \exp(\varepsilon|x|).$$

Proof of the theorem follows from two lemmas.

Lemma 5.4. Let conditions of Theorem 5.7 be satisfied. Then for any $\delta \in (0, \delta_0)$ there exists such $\varepsilon > 0$, and conversely for any $\varepsilon \in (0, \varepsilon_0)$ there exists such $\delta > 0$, that

$$\mathsf{E} \exp(\delta\tau) \leq \exp(\varepsilon|X_0|).$$

Lemma 5.5. Let conditions of Theorem 5.7 be satisfied. Then for any $\varepsilon \in (0, \varepsilon_0)$ and any $n \geq 0$

$$\mathsf{E} \exp(\varepsilon|X_n|) \leq C_\varepsilon \exp(\varepsilon|X_0|),$$

and, moreover, $C_\varepsilon \to 0$ as $\varepsilon \to 0$.

Proof of Lemma 5.4. For $n < \tau$

$$\mathsf{E}(\exp(\delta(n+1) + \varepsilon|X_{n+1}|)|X_n) = \mathsf{E}(\exp(\delta(n+1) + \varepsilon|X_n + a(X_n) + b(X_n)W_n|)|X_n).$$

Let us denote

$$z_n = X_n + a(X_n), \quad \eta_n = b(X_n)W_n, \quad e_z = (z/|z|)\,\mathrm{I}(z \neq 0) + (1, 0, \ldots, 0)^*\,\mathrm{I}(z = 0).$$

Then

$$|X_n + a(X_n) + b(X_n)W_n| = |z_n + \eta_n| = \langle z_n + \eta_n, e_{z_n + \eta_n}\rangle.$$

Thus,

$$\begin{aligned} &\mathsf{E}(\exp(\varepsilon|X_n + a(X_n) + b(X_n)W_n|)|X_n) \\ &\quad = \mathsf{E}(\exp(\varepsilon\langle z_n + \eta_n, e_{z_n+\eta_n}\rangle)|X_n) \\ &\quad = \mathsf{E}(\exp(\varepsilon\langle z_n + \eta_n, e_{z_n}\rangle + \varepsilon\langle z_n + \eta_n, e_{z_n+\eta_n} - e_{z_n}\rangle)\,\mathrm{I}(|\eta_n| \leq \rho)|X_n) \\ &\quad\quad + \mathsf{E}(\exp(\varepsilon\langle z_n + \eta_n, e_{z_n+\eta_n}\rangle)\,\mathrm{I}(|\eta_n| > \rho)|X_n) \equiv A_1 + A_2. \end{aligned}$$

Let us choose such $\rho > 0$ that $\mathsf{P}(|\eta_n| > \rho/2|n < \tau) \leq \Delta$, $\Delta > 0$. If R is large enough then we obtain for $|\eta_n| \leq \rho/2$ and $n < \tau$

$$|e_{z_n+\eta_n} - e_{z_n}| \leq \frac{\rho}{|z_n^{\varsigma}|},$$

hence, for $n < \tau$

$$\mathsf{P}(|e_{z_n+\eta_n} - e_{z_n}| > \rho/|z_n||X_n) \leq \Delta.$$

Let us estimate A_2 for $n < \tau$:

$$\begin{aligned} A_2 &\leq \exp(\varepsilon|z_n|)\,\mathsf{E}(\exp(\varepsilon|\eta_n|)\,\mathrm{I}(|e_{z_n+\eta_n} - e_{z_n}| > \rho/|z_n|)|X_n) \\ &\leq \exp(\varepsilon(|X_n| - \nu))(\mathsf{E}\exp(2\varepsilon|\eta_n|))^{1/2}\,\mathsf{P}(|e_{z_n+\eta_n} - e_{z_n}| > \rho/|z_n||X_n)^{1/2} \\ &\leq \nu^{-1/2}\exp(\varepsilon|X_n|)\,\mathsf{P}(|e_{z_n+\eta_n} - e_{z_n}| > \rho/|z_n||X_n)^{1/2}. \end{aligned}$$

Thus,

$$A_2 \leq \nu^{-1/2}\Delta^{1/2}\exp(\varepsilon|X_n|),$$

and the factor $\nu^{-1/2}\Delta^{1/2} > 0$ here may be done arbitrary small by choosing of ρ and R.

Let us estimate A_1 for $n < \tau$:

$$\begin{aligned}
A_1 &\le \mathsf{E}(\exp(\varepsilon\langle z_n + \eta_n, e_{z_n}\rangle + 2\varepsilon(|z_n| + \rho)(\rho/|z_n|)^2)|X_n) \\
&= \mathsf{E}(\exp(\varepsilon\langle z_n + \eta_n, e_{z_n}\rangle + 2\varepsilon(\rho^2|z_n|^{-1} + \rho^3|z_n|^{-2}))|X_n) \\
&= \exp(\varepsilon|z_n| + 2\varepsilon\rho^2R^{-1} + 2\varepsilon\rho^3R^{-2})\,\mathsf{E}(\exp(\varepsilon\langle \eta_n, e_{z_n}\rangle)|X_n) \\
&\le \exp(\varepsilon|X_n| - \varepsilon(\nu - 2\rho^2R^{-1} - 2\rho^3R^{-2})\,\mathsf{E}(\exp(\varepsilon\langle \eta_n, e_{z_n}\rangle)|X_n).
\end{aligned}$$

Now, let us expand the value $\exp(\varepsilon\langle \eta_n, e_{z_n}\rangle)$. If $(\varepsilon\nu^{-1})^{3/4} \le \varkappa$ then we have

$$\begin{aligned}
\mathsf{E}(\exp(\varepsilon\langle \eta_n, e_{z_n}\rangle)|X_n) &= \mathsf{E}\left(\sum_{k=0}^{\infty}\varepsilon^k\langle \eta_n, e_{z_n}\rangle^k/k!|X_n\right) \\
&\le 1 + \mathsf{E}\left(\sum_{k=2}^{\infty}\varepsilon^k|\eta_n|^k/k!|X_n\right) \\
&\le 1 + \mathsf{E}\sum_{k=2}^{\infty}\varepsilon^k\nu^{-k}|W_n|^k/k! \\
&= 1 + \mathsf{E}\sum_{k=2}^{\infty}(\varepsilon\nu^{-1})^{3k/4}\,\mathsf{E}(\varepsilon\nu^{-1})^{k/4}|W_n|^k/k! \\
&\le 1 + (\varepsilon\nu^{-1})^{3/2}\,\mathsf{E}\sum_{k=2}^{\infty}(\varepsilon\nu^{-1})^{k/4}|W_n|^k/k! \\
&\le 1 + (\varepsilon\nu^{-1})^{3/2}\,\mathsf{E}\exp((\varepsilon\nu^{-1})^{1/4}|W_n|) \\
&\le 1 + C\nu^{-3/2}\varepsilon^{3/2}.
\end{aligned}$$

Thus,

$$A_1 \le \exp(\varepsilon|X_n|)\exp(-\varepsilon(\nu - 2\rho^2R^{-1} - 2\rho^3R^{-2}))(1 + C\varepsilon^{3/2}),$$

where $\nu^{-3/2}$ is included in C. The second exp here does not exceed $1 - \varepsilon\nu/2$ if R is large enough. Then

$$A_1 \le \exp(\varepsilon|X_n|)(1 - \varepsilon\nu/2)(1 + C\varepsilon^{3/2}).$$

If ε is small enough then

$$A_1 \le \exp(\varepsilon|X_n|)(1 - \varepsilon\nu/3).$$

Thus,

$$A_1 + A_2 \le \exp(\varepsilon|X_n|)(1 - \varepsilon\nu/3 + \nu^{-1/2}\Delta^{1/2}).$$

At last, by choosing the values Δ, ρ and R we obtain

$$A_1 + A_2 \le \exp(\varepsilon|X_n|)(1 - \varepsilon\nu/6).$$

Hence, for $n < \tau$

$$\begin{aligned}
&\mathsf{E}(\exp(\delta(n+1) + \varepsilon|X_{n+1}|)|X_n) - \exp(\delta n + \varepsilon|X_n|) \\
&\qquad\le \exp(\delta n + \varepsilon|X_n|)\exp(\delta)(1 - \varepsilon\nu/6).
\end{aligned}$$

So for any $\delta \in (0, \delta_0)$ there exists such $\varepsilon > 0$, and, conversely, for any $\varepsilon \in (0, \varepsilon_0)$ there exists such $\delta > 0$, that

$$\mathsf{E}\,(\exp(\delta(n+1) + \varepsilon|X_{n+1}|)|X_n) - \exp(\delta n + \varepsilon|X_n|) \le -\varepsilon\nu/12 \exp(\delta n + \varepsilon|X_n|).$$

Thus, for any $n \ge 0$

$$\mathsf{E}\exp(\delta(n \wedge \tau)) \le \exp(\varepsilon(|X_0|),$$

and B. Levi's theorem on the monotonous convergence implies the estimate together with the finiteness of the moment τ. Lemma 5.4 is proved.

Proof of Lemma 5.5. Let us consider the process $g_n = \exp(\varepsilon|X_n|)$, $n \ge 0$. We have for R fixed

$$\mathsf{E}\,((g_{n+1} - g_n)\,\mathrm{I}\,(|X_n| \le R)|X_n) \le K_\varepsilon, \qquad K_\varepsilon > 0.$$

At the same time, if R is large enough then

$$\mathsf{E}\,((g_{n+1} - g_n)\,\mathrm{I}\,(|X_n| > R)|X_n) \le -\varepsilon\nu^{-1}g_n/12.$$

Thus,

and for any $n \ge 0$

$$\mathsf{E}g_n \le 12K_\varepsilon\varepsilon^{-1}\nu.$$

In the other words,

$$\mathsf{E}\exp(\varepsilon|X_n|) \le 12K_\varepsilon\varepsilon^{-1}\nu.$$

Hence, by virtue of the Hölder inequality for any $k \ge 1$

$$\mathsf{E}\exp(\varepsilon k^{-1}|X_n|) \le C_\varepsilon \equiv (12K_\varepsilon\varepsilon^{-1}\nu)^{1/k}.$$

If k here is large enough then C_ε may be chosen in an arbitrary neighbourhood of 1. Lemma 5.5 is proved.

Comments

Conditions of strong mixing, uniformly strong mixing et al. are often used in limit theorems. On the other hand, there are not so many classes of processes for which the conditions, including any bounds for coefficients, may be checked. There are bounds for complete regularity coefficients for Gaussian stationary processes obtained by Ibragimov (1961, 1962) and Kolmogorov and Rozanov (1960), see also Rozanov (1967). Davydov (1973) obtained the exponential bounds for the ergodic Markov chain uniformly strong mixing coefficient under Doeblin-type conditions. Various properties of mixing for Markov chains with Doeblin-type condition and some stationary processes were studied by Bloom, Hanson and Koopmans (1963), Kesten and O'Brien (1976), Bradley (1980, 1983).

The chapter is based on Veretennikov (1987, 1991a).

Weaker conditions in semimartingale terms under which the central limit theorem is known may be found in Liptser and Shiryaev (1977). Nevertheless, the applicability of these conditions to any class of processes is a problem to be investigated.

Bounds of mixing coefficients are used in stochastic approximation theory, see Kushner and Clark (1978). In Benveniste, Metivier and Priouret (1987) some methods of some rate of mixing on Lipschitz class estimation is introduced. These estimates are weaker than usual mixing but are sufficient for some stochastic approximation problems. We introduce in some sense the alternative method based on convergence in variance. Our results are thus stronger but the method requires more restrictive conditions on transition probabilities (some local variant of Doeblin-type conditions).

We exploit the popular coupling method. One may read the modern coupling method in Nummelin (1984). Usually the coupling method is used to prove the convergence in variance or to find the rate of such convergence. We have not seen the applications of this method to mixing coefficients although these problems are not very different. The beginning of the coupling method is considered to be in Dobrushin (1971) and Vaserstein (1969). Indeed, this method is the extension of Kolmogorov's and Doeblin's ideas. It may also be thought of as being a variant of the known unique probability space method. Nevertheless, the special name for this method is of course correct.

This approach works in some non-Markov cases as well. Some applications to mixing rate for stationary processes will appear in Veretennikov, Proc. VI Soviet–Japan Symp. on Probability Theory and Mathematical Statistics (USSR, Kiev, August 1991).

Chapter 6

The averaging principle for some stochastic recursive schemes with state dependent noise

Setting. Main results

In the previous section we have made all the preparations necessary to prove the averaging principle for the recursive procedure:

$$\theta_{n+1} = \theta_n + \gamma f(\theta_n, Y_{n+1}), \tag{6.1}$$

$$Y_{n+1} = Y_n + a(\theta_n, Y_n) + b(\theta_n, Y_n)W_n, \qquad n \geq 0, \tag{6.2}$$

where $W = (W_n)$ is the sequence of i.i.d. random vectors with values in E^m, $\theta_n \in E^p$, $Y_n \in E^d$ are random vectors, f, a, b are measurable bounded functions, $f(\theta, y) \in E^p$, $a(\theta, y) \in E^d$, $b(\theta, y)$ is an $(m \times d)$-matrix, $\gamma = \text{const} \ll 1$, θ_0, Y_0 are some non-random initial conditions.

Before getting into details it may be useful to make a couple of heuristic remarks. Given θ, define a process $Y^\theta = (Y_n^\theta)$, $n \geq 0$, with

$$Y_{n+1}^\theta = Y_n^\theta + a(\theta, Y_n^\theta) + b(\theta, Y_n^\theta)W_n, \tag{6.3}$$

and assume that Y^θ has an invariant measure μ^θ. Define a function

$$h(\theta) = \int f(\theta, y)\mu^\theta(\mathrm{d}y).$$

For small γ, $\theta = (\theta_n)$, $n \geq 0$, is a 'slow' process comparably with the 'fast' process (Y_n), $n \geq 0$. We introduce a new 'time' $t_n = \gamma n$, denote $m(t) = \max(n\colon t_n \leq t)$ and define $\theta(t)$ as a piecewise constant version of θ_n, i.e., $\theta(t) = \theta_n$ as $t_n \leq t < t_{n+1}$. Then we can write (6.1) in the following form:

$$\begin{aligned}\theta(t) &= \theta_0 + \gamma \sum_{k=1}^{m(t)} f(\theta_{k-1}, Y_k) \\ &= \theta_0 + \gamma \sum_{k=1}^{m(t)} h(\theta_{k-1}) + \gamma \sum_{k=1}^{m(t)} [f(\theta_{k-1}, Y_k) - h(\theta_{k-1})]\end{aligned}$$

or as continuous time version

$$\theta(t) = \theta_0 + \int_0^t h(\theta(s))\,\mathrm{d}s + R(t),$$

with

$$R(t) = \gamma \sum_{k=1}^{m(t)} [f(\theta_{k-1}, Y_k) - h(\theta_{k-1})] - \int_{t_{m(t)}}^t h(\theta(s))\,\mathrm{d}s. \tag{6.4}$$

This equation can be considered as the result of random perturbation of the ordinary differential equation

$$\frac{\mathrm{d}\bar{\theta}(t)}{\mathrm{d}t} = h(\bar{\theta}(t)), \qquad \bar{\theta}(0) = \theta_0.$$

If the process Y^θ has a good mixing property and $a(\theta, y)$, $b(\theta, y)$ depend 'weakly' on the 'slow' process θ, one may expect that as $\gamma \to 0$, the following ergodic type result will take place: for any $c > 0$,

$$\lim_{\gamma \to 0} \mathsf{P}\left(\sup_{0 \le t \le T} |\theta(t) - \bar{\theta}(t)| > c\right) = 0.$$

We will refer to this property as the averaging principle.

We now formulate the assumptions we need to impose on the process Y^θ to obtain the averaging principle.

A. There exists $\nu > 0$ such that for all θ

$$\sup_{\theta, y} (|a(\theta, y)| + \|b(\theta, y)\|) \le \nu^{-1}.$$

B. There exists $R > 0$ such that for all θ and $|y| > R$

$$(a(\theta, y), y) \le -\nu |y|.$$

C. W is the sequence of i.i.d. random vectors with $\mathsf{E} W_n = 0$, and for some $\alpha > 0$

$$\mathsf{E} \exp(\alpha W_1) < \infty.$$

D. Let P_z be a distribution of the random value $z + b(\theta, y) W_n$ and $q_{z_1, z_2}(z)$ be a density of the absolutely continuous w.r.t P_{z_2} component of the measure P_{z_1}. There exist $R_1 > R$ and $\nu(R_1) > 0$ such that

$$\inf_{|z_1|, |z_2| \le R_1} \int_{|z| \le R_1} 1 \wedge q_{z_1, z_2}(z) P_{z_2}(\mathrm{d}z) > \nu(R_1).$$

One can see that if conditions A–D hold, then Y^θ satisfies the conditions of Theorem 5.7 of the previous chapter. Hence there exist C, λ and $\varepsilon > 0$ which depend on α, ν such that for all $n \ge 0$

$$\alpha(n) \le \beta(n) \le C \exp(\varepsilon |Y_0|) \exp(-\lambda n), \tag{6.5}$$

the invariant measure μ^θ has the property

$$\int \exp(\varepsilon |y|) \, \mathrm{d}\mu^\theta \le \infty$$

and there exists $K(\mu^\theta)$ such that for all Y_0 and n

$$\operatorname{var}(P_n^\theta - \mu^\theta) \le K(\mu^\theta) \exp(\varepsilon |Y_0|) \exp(-\lambda n), \tag{6.6}$$

where P_n^θ is the distribution of Y_n^θ.

We assume that $h(\theta) = \int f(\theta,y)\mu^\theta(dy)$ and f in (6.1) satisfy the following conditions:

$$|h(\theta) - h(\theta')| \leq K_1|\theta - \theta'|, \qquad K_1 > 0; \tag{6.7}$$

f is bounded and

$$|f(\theta,y) - f(\theta',y')| \leq K_2(|\theta - \theta'| + |y - y'|), \qquad K_2 > 0. \tag{6.8}$$

Theorem 6.1. Let conditions A–D and (6.7), (6.8) hold. Then for $T > 0$,

$$\lim_{\gamma\to 0} \mathsf{E} \sup_{0\leq t\leq T} |\theta(t) - \bar{\theta}(t)| = 0.$$

Proof of the main result
We need the following corollary of Theorem 5.7.

Lemma 6.1. Let $g(y)$, $y \in E^d$, be measurable and bounded function. Then

$$\left|\mathsf{E}\left(n^{-1}\sum_{i=k+1}^{k+n} g(Y_i^\theta) - \int g(y)\mu^\theta(dy)\right)\right| \leq \|g\|_{\mathrm{B}} K n^{-1} \exp(\varepsilon|Y_0|)\lambda^{-1}\exp(-\lambda k), \tag{6.9}$$

where $\|\cdot\|_{\mathrm{B}}$ is the uniform norm.

Proof. It follows form the conditions A–D that for some $\varepsilon > 0$ we have

$$\int \exp(\varepsilon|y|)\, d\mu^\theta < \infty$$

for any θ. Thus, in virtue of (6.6), we get

$$\begin{aligned}\operatorname{var}\left(n^{-1}\sum_{i=k+1}^{k+n} P_i^\theta - \mu^\theta\right) &\leq K n^{-1}\exp(\varepsilon|Y_0|)\sum_{i=k+1}^{k+n}\exp(-\lambda i)\\ &\leq K n^{-1}\exp(\varepsilon|Y_0|)\int_k^\infty \exp(-\lambda s)\, ds\\ &= K\lambda^{-1}n^{-1}\exp(\varepsilon|Y_0|)\exp(-\lambda k)\end{aligned}$$

and so

$$\begin{aligned}\left|\mathsf{E}\left(n^{-1}\sum_{i=k+1}^{k+n} g(Y_i^\theta) - \int g(y)\mu^\theta(dy)\right)\right| &\leq \|g\|_{\mathrm{B}}\operatorname{var}\left(n^{-1}\sum_{i=k+1}^{k+n} P_i^\theta - \mu^\theta\right)\\ &\leq K\lambda^{-1}n^{-1}\exp(\varepsilon|Y_0|)\exp(-\lambda k).\end{aligned}$$

The lemma is proved.

Proof of Theorem 6.1. We follow the wellknown scheme (see, for example, Wentzell and Freidlin (1984)).

Step 1. In view of the Lipschitz condition (6.7), we have

$$\mathsf{E}\sup_{0\le s\le t}|\theta(s)-\bar\theta(s)| \le K_1\int_0^t \mathsf{E}\sup_{0\le r\le s}|\theta(r)-\bar\theta(r)|\,\mathrm{d}s + \mathsf{E}\sup_{0\le s\le t}|R(s)|,$$

with $R(t)$, defined by (6.4). Thus, by Gronwall's inequality,

$$\mathsf{E}\sup_{0\le s\le t}|\theta(s)-\bar\theta(s)| \le \mathsf{E}\sup_{0\le s\le t}|R(s)|\exp(K_1 t). \tag{6.10}$$

Step 2. Since $0\le t-t_{m(t)}\le\gamma$, then

$$\left|\int_{t_{m(t)}}^t h(s)\,\mathrm{d}s\right| \le \gamma\|h\|_{\mathbf{B}}.$$

Thus, in view of (6.10) and (6.4), we only have to prove that

$$\lim_{\gamma\to0}\gamma\,\mathsf{E}\sup_{0\le s\le t}\left|\sum_{k=1}^{m(s)}[f(\theta_{k-1},Y_k)-h(\theta_{k-1})]\right| = 0. \tag{6.11}$$

Pick Δ be a multiple of γ and such that $\Delta\to0$ and $\gamma/\Delta\to0$ as $\gamma\to0$. Denote $p(t)=[t/\Delta]$. Then

$$\begin{aligned}
\gamma\sum_{k=1}^{m(s)}[f(\theta_{k-1},Y_k)-h(\theta_{k-1})] &= \gamma\sum_{k=0}^{p(t)-1}\sum_{i=m(k\Delta)}^{m(k\Delta+\Delta)-1}[f(\theta_i,Y_{i+1})-h(\theta_i)]\\
&\quad+\gamma\sum_{i=m(p(t))}^{m(t)-1}[f(\theta_i,Y_{i+1})-h(\theta_i)]\\
&= \gamma\sum_{k=0}^{p(t)-1}\sum_{i=m(k\Delta)}^{m(k\Delta+\Delta)-1}[f(\theta_{m(k\Delta)},Y_{i+1}^{\theta_{m(k\Delta)}})-h(\theta_{m(k\Delta)})]\\
&\quad+\gamma\sum_{k=0}^{p(t)-1}\sum_{i=m(k\Delta)}^{m(k\Delta+\Delta)-1}[f(\theta_i,Y_{i+1})-f(\theta_{m(k\Delta)},Y_{i+1}^{\theta_{m(k\Delta)}})]\\
&\quad+\gamma\sum_{k=0}^{p(t)-1}\sum_{i=m(k\Delta)}^{m(k\Delta+\Delta)-1}[h(\theta_{m(k\Delta)})-h(\theta_i)]\\
&\quad+\gamma\sum_{i=m(p(t)\Delta)}^{m(t)-1}[f(\theta_i,Y_{i+1})-h(\theta_i)]\\
&\equiv I_t^1+I_t^2+I_t^3+I_t^4.
\end{aligned}$$

Step 3. The choice of Δ and $p(t)$ implies

$$\lim_{\gamma\to0}\mathsf{E}\sup_{0\le t\le T}|I_t^4| = 0.$$

Keeping in mind the last argument, from Lipschitz conditions (6.7) and (6.8) one could easily get

$$\lim_{\gamma\to 0} \mathsf{E} \sup_{0\le t\le T} |I_t^2| = 0,$$

and

$$\lim_{\gamma\to 0} \mathsf{E} \sup_{0\le t\le T} |I_t^3| = 0.$$

Thus, it remains to estimate I_t^1.

Step 4. Pick ψ be the multiple of γ, such that $\psi + \gamma/\psi + \psi/\Delta \to 0$ as $\gamma \to 0$ and consider

$$\begin{aligned}\gamma^{-1} I_t^1 &= \gamma \sum_{k=0}^{p(t)-1} \left(\sum_{i=m(k\Delta)}^{m(k\Delta+\psi)-1} + \sum_{i=m(k\Delta+\psi)}^{m(k\Delta+\Delta)-1} \right) [f(\theta_{m(k\Delta)}, Y_{i+1}^{\theta_{m(k\Delta)}}) - h(\theta_{m(k\Delta)})] \\ &\equiv J_t^1 + J_t^2.\end{aligned}$$

Evidently,

$$|J_t^1| = \left| \sum_{k=0}^{p(t)-1} \sum_{i=m(k\Delta)}^{m(k\Delta+\psi)-1} [f(\theta_{m(k\Delta)}, Y_{i+1}^{\theta_{m(k\Delta)}}) - h(\theta_{m(k\Delta)})] \right| \le C\psi/\gamma,$$

and by the choice of Δ and ψ as $\gamma \to 0$

$$\gamma(t/\Delta) C(\psi/\gamma) = Ct\gamma/\Delta \to 0.$$

Step 5. The final estimation is the main in our proof. We apply estimation (6.9) of Lemma 6.1 with $m(k\Delta + \psi) \le i \le m(k\Delta + \Delta) - 1$ to get

$$\begin{aligned}\mathsf{E} \sup_{0\le t\le T} |J_t^2| &\le \gamma \sum_{k=0}^{p(t)-1} \left| \mathsf{E} \sum_{i=m(k\Delta+\psi)}^{m(k\Delta+\Delta)-1} [f(\theta_{m(k\Delta)}, Y_{i+1}^{\theta_{m(k\Delta)}}) - h(\theta_{m(k\Delta)})] \right| \\ &\quad \times |\theta_{m(k\Delta)}, Y_{m(k\Delta)})| + C\Delta \\ &\le C\Delta + \gamma \mathsf{E} \sum_{k=0}^{p(t)-1} \|f\|_{\mathrm{B}} K \lambda^{-1} \exp(-\lambda\psi/\gamma) \exp(\varepsilon |Y_{m(k\Delta)}|) \\ &= C\Delta + \gamma \|f\|_{\mathrm{B}} K \lambda^{-1} \exp(-\lambda\psi/\gamma) \sum_{k=0}^{p(t)-1} \mathsf{E} \exp(\varepsilon |Y_{m(k\Delta)}|).\end{aligned}$$

Since by Lemma 5.5 of the previous chapter,

$$\mathsf{E} \exp(\varepsilon |Y_{m(k\Delta)}|) \le C \exp(\varepsilon |Y_0|)$$

we get

$$\mathsf{E} \sup_{0\le t\le T} |J_t^2| \le C\Delta + C\gamma T \Delta^{-1} \exp(\varepsilon |Y_0| - \lambda\psi/\gamma) \to 0,$$

as $\gamma \to 0$. This and the estimations of steps 3 and 4 imply (6.11). The theorem is proved.

Comments

There is a sizeable literature concerning the averaging principle. Continuous time processes were studied by Khasminskii (1966) and Freidlin (1978). Discrete time processes in the context of a convergence of stochastic recursive algorithms were considered by many authorh (see Derevitskii and Fradkov (1981), Ljung (1977), Kushner and Clark (1978) and Benveniste, Metivier and Priouret (1987) for detailed discussion and references). The assumptions of Theorem 6.1 are more then enough for proving the averaging principle. Nevertheless these assumptions are natural for large deviation analysis. The exposition of the chapter is based on the paper by Veretennikov and Gulinsky (1990).

Chapter 7

Normal deviations

Let us study the problem of normal deviations in the averaging principle. We consider the simplest case only when the fast process does not depend on the slow one. Such a situation was studied in Khasminskii (1966) and Freidlin and Wentzell (1984), Chapter 7. Consider the system

$$\begin{aligned} \vartheta_{n+1} &= \vartheta_n + \gamma f(\vartheta_n, Y_n), \qquad \vartheta_0 = \vartheta, \\ Y_{n+1} &= Y_n + a(Y_n) + b(Y_n)W_n, \qquad Y_0 = Y, \qquad n \le T/\gamma. \end{aligned} \tag{7.1}$$

Let us denote $\tilde{\vartheta}_t = \vartheta_{[t/\gamma]\gamma}$, $\tilde{y}_t = Y_{[t/\gamma]\gamma}$.

Theorem 7.1. Let following conditions hold:

(1) $f \in C_b^2$;
(2) a and b are bounded;
(3) $(D_K) + (A_{K,n}) + (B_{K,n})$ for some K and some $n > 10$.

Then the process $\zeta_t^\gamma = \gamma^{-1/2}(\tilde{\vartheta}_t - \bar{\vartheta}_t)$ converges weakly on $[0, \infty)$ as $\gamma \to 0$ to Markov–Gaussian process ζ_t^0, which is a solution of the linear differential equation

$$\mathrm{d}\zeta_t^0 = B(\bar{\vartheta}_t)\zeta_t^0\,\mathrm{d}t + \mathrm{d}w_t^0, \qquad \zeta_0^0 = 0, \tag{7.2}$$

where w^0 is a continuous Gaussian process with independent increments, zero expectation and covariance matrix $R(t) = (R^{ki}(t))$,

$$R^{ki}(t) = \mathsf{E}w_t^{0,k}w_t^{0,i} = \int_0^t A^{ki}(\bar{\vartheta}_s)\,\mathrm{d}s,$$

$$A^{ki}(x) = \lim T^{-1}\int_{t_0}^{t_0+T}\int_{t_0}^{t_0+T} A^{ki}(x,s,t)\,\mathrm{d}t\,\mathrm{d}s, \tag{7.3}$$

$$A^{ki}(x,s,t) = \mathsf{E}[f^k(x,\tilde{y}_s) - \mathsf{E}f^k(x,\tilde{y}_s)][f^i(x,\tilde{y}_s) - \mathsf{E}f^i(x,\tilde{y}_s)],$$

$$B(x) = (B_j^i(x)) = (\partial\bar{f}^i/\partial x^j(x)),$$

$$\bar{f}(x) = \lim T^{-1}\int_{t_0}^{t_0+T} \mathsf{E}f(x,\tilde{y}_s)\,\mathrm{d}s \qquad (= \mathsf{E}^{\mathrm{inv}} f(x,\tilde{y}_s)). \tag{7.4}$$

The existence of the limits in (7.3)–(7.4) is included in the assertion of the theorem. Note that $\bar{f} \in C_b^2$ as well as f for the stationary measure of the fast process does not depend on x.

Remark 7.1. Linear equation (7.2) may be solved in implicit form, see, for example, Liptser and Shiryaev (1977).

Remark 7.2. We follow the proof in Freidlin and Wentzell (1984), though the use of modern compactness conditions in C may weaken the stability conditions for the fast process.

Proof of Theorem 7.1. We reduce the proof to Theorem 7.3.1 from Freidlin and Wentzell (1984). For this aim we remind its conditions:

(i) $b', b'' \in C_b$;

(ii) the process Y is pathwise continuous a.e. and

$$\sup_{x,n} \mathsf{E}|b(x, Y_n)| < \infty, \qquad \int_0^\infty \tau\alpha(\tau)^{1/5}\,\mathrm{d}\tau < \infty,$$

where α is the strong mixing coefficient of Y;

(iii) the following limits exist uniform in x, t_0:

$$\lim_{T\to\infty} T^{-1} \int_{t_0}^{t_0+T} \mathsf{E}f(x, Y_t)\,\mathrm{d}t = \bar{f}(x),$$

$$\lim_{T\to\infty} T^{-1} \int_{t_0}^{t_0+T}\int_0^{+T} A^{ki}(x, s, t)\,\mathrm{d}s\,\mathrm{d}t = A^{ki}(x),$$

where $A^{ki}(x)$ are from the assertion of the theorem;

(iv) for any $\tau \in [0, T_0]$ and some $C > 0$

$$\left|\int_0^\tau [\mathsf{E}f(\bar{x}_t, Y_{t/\varepsilon})\,\mathrm{d}t - \bar{f}(\bar{x}_s)]\right| \mathrm{d}s \le C\varepsilon,$$

$$\max_{k,i}\left|\int_0^\tau \left[\mathsf{E}\frac{\partial f^k}{\partial x^i}(\bar{x}_t, Y_{t/\varepsilon})\,\mathrm{d}t - \frac{\partial \bar{f}^k}{\partial x^i}(\bar{x}_s)\right]\right| \mathrm{d}s \le C\varepsilon.$$

Under these conditions the same assertion is stated as in Theorem 7.1 for the continuous time differential system

$$\mathrm{d}Z_t^\gamma = \gamma f(Z_t^\gamma, Y_t)\,\mathrm{d}t, \qquad Z_0^\gamma = x,$$

but by virtue of our condition (1) this is equivalent to our assertion. (We define the process Y_t for any t by the equality $Y_t = Y_{[t]}$.)

Condition (ii) is satisfied by virtue of conditions and results from Chapter 5.

The first limit in (iii) exists by virtue of mixing results from Chapter 5 as well.

The second limit in (iii) exists by virtue of similar reasons. Indeed, by virtue of Theorem 5.6 the difference of the value $\mathsf{E}f^k(x, \tilde{y}_s)$ and the expectation $\mathsf{E}^{\mathrm{inv}} f^k(x, \tilde{y}_s)$ w.r.t. the stationary measure does not exceed $C(1+s)^{-10}$, and so is $\mathsf{E}f^i(x, \tilde{y}_t)$. Thus, the limit does exists and it is equal to the integral of the covariance matrix of Y_t in the stationary regime for x fixed.

The first inequality in (iv) holds also by virtue of theorems on mixing.

The second inequality follows from similar considerations and from the equality

$$\nabla \bar{b}(x) = \int \nabla_x b(x, y)\, \mu(\mathrm{d}y),$$

where μ is the stationary measure of the fast process. The last equality follows from the smoothness conditions.

Thus, all conditions of Theorem 7.3.1 in Freidlin and Wentzell (1984) are satisfied. The theorem is proved.

Comments

Theorem 7.1 is the consequence of Theorem 7.3.1 from Freidlin and Wentzell (1984) for all conditions of the latter are derived from some more concrete conditions which are very easily checked. It is a functional central limit type theorem. There is analogous result in the case when the fast process depends on the slow one, see Benveniste, Metivier and Priouret (1987). Our method also works in that case but the work is not complete.

Chapter 8

Large deviations for Markov processes

In the beginning of the chapter the introduction to the theory of large deviations is presented based on so-called 'property F' due to Freidlin and on the Frobenius-type theorem on positive operators. The main example in this chapter is a Markov process which satisfies the local Doeblin-type condition.

8.1. GÄRTNER'S THEOREM

We follow Section 1.5 of Freidlin and Wentzell (1984). Let $G(\alpha)$, $\alpha \in E^l$, be a convex, lower semicontinuous function with values in $(-\infty, \infty]$ which does not equal ∞ identically. (If function G is convex then one may check its lower semicontinuity $G(\alpha) \le \liminf_{\alpha' \to \alpha} G(a)$ for $\alpha \in \partial\{\alpha' \colon G(\alpha') < \infty\}$ only.) Let $L(\beta)$, $\beta \in E^l$, be a Fenchel–Legendre transformation of function G:

$$L(\beta) = \sup_{\alpha} (\alpha\beta - G(\beta)).$$

Function L is also convex, lower semicontinuous, with values in $(-\infty, \infty]$ and does not equal ∞. The equality holds

$$G(\alpha) = \sup_{\alpha} (\alpha\beta - L(\alpha))$$

(see Rockafellar (1970), Theorem 12.2). For any point α_0 internal for the set $\{\alpha \colon G(\alpha) < \infty\}$ w.r.t. the linear hall there exists such vector β_0 (it is called the subgradient of G in α_0), so that for any α

$$G(\alpha) \ge G(\alpha_0) + (\alpha - \alpha_0)\beta_0;$$

it means also that

$$G(\alpha) = \alpha_0\beta_0 - L(\beta_0)$$

(see Rockafellar (1970), Theorem 23.4). Function $L(\beta)$ tends to $+\infty$ as $\beta \to \infty$ if and only if $G(\alpha) < \infty$ in some neighborhood of zero.

Let μ^h be a family of probability measures in E^l. Let

$$G^h(\alpha) := \ln \int \exp(\alpha x)\, \mu^h(\mathrm{d}x),$$

if the integral is finite, and $+\infty$ otherwise.

Function G^h is convex: indeed, for $c \in (0, 1)$ by virtue of the Hölder inequality

$$\begin{aligned} G^h(c\alpha_1 + (1-c)\alpha_2) &= \ln \int \exp(c\alpha_1 x) \exp((1-c)\alpha_2 x)\, \mu^h(\mathrm{d}x) \\ &\le \ln \left[\left(\int \exp(\alpha_1 x)\, \mu^h(\mathrm{d}x) \right)^c \left(\int \exp(\alpha_2 x)\, \mu^h(\mathrm{d}x) \right)^{1-c} \right] \\ &= cG^h(\alpha_1) + (1-c)\, G^h(\alpha_2). \end{aligned}$$

Function G^h takes its values in $(-\infty, +\infty]$, it is lower semicontinuous (by virtue of Fatou lemma) and does not equal $+\infty$ identically as $G^h(0) = 0$.

Let $\lambda(h)$ be a function with values in R^1, which tends to $+\infty$ as $h \to 0$. Assume that for any α there exists the limit

$$G(\alpha) = \lim_{h\to 0} \lambda(h)^{-1} G^h(\lambda(h)\alpha). \tag{8.1}$$

Function $G(\alpha)$ is convex and $G(\alpha) = 0$. Suppose that $G(\alpha)$ is lower semicontinuous (as we have mentioned this is the condition on its behaviour on the set $\partial\{\alpha\colon G(\alpha) < \infty\}$), G does not equal $-\infty$, and G is finite in some neighbourhood of zero. Let $L(\beta) = \sup(\alpha\beta - G(\alpha))$.

Theorem 8.1 (Gärtner). For any $\delta > 0$, $\gamma > 0$, $s > 0$ there exists such $h_0 > 0$ that for any $h \le h_0$

$$\mu^h(y\colon \mathrm{P}\rho(y, \Phi(s)) \ge \delta) \le \exp(-\lambda(s - \gamma)), \tag{8.2}$$

where $\Phi(s) = (\beta\colon L(\beta) \le s)$ is a compact.

Proof. The compactness of $\Phi(s)$ follows from the lower semicontinuity of L and the property $\lim_{\beta\to\infty} L(\beta) = +\infty$. The following representation holds:

$$\Phi(s) = \bigcap_{\alpha} (\beta\colon \alpha\beta - G(\alpha) \le s).$$

Let us consider the boundary of δ-neighbourhood of $\Phi(s)$:

$$\partial\Phi_\delta(s) = \partial(z\colon \rho(z, \Phi(s)) \le \delta) = (z\colon \rho(z, \Phi(s)) = \delta).$$

This set is compact as the boundary of a compact. For any $z \in \partial\Phi_\delta(s)$ there exists such α that

$$\alpha z - G(\alpha) > s,$$

then open halfspaces $(z\colon \alpha z - G(\alpha) > s)$ cover the compact $\partial\Phi_\delta(s)$. Let us choose the finite subcovering with some $\alpha_1, \dots, \alpha_N$. Then the convex polyhedron

$$\bigcap_{i=1}^{N} (z\colon \alpha_i z - G(\alpha_i) \le s)$$

contains $\Phi(s)$ and does not intersect with $\partial\Phi_\delta(s)$, that is,

$$\bigcup_{i=1}^{N} (z\colon \alpha_i z - G(\alpha_i) > s) \supset E^l \backslash \Phi_\delta(s).$$

Thus, by virtue of exponential Chebyshov's inequality

$$\begin{aligned}
\mu^h(y\colon \rho(y, \Phi(s)) \ge \delta) &\le \mu^h\left(\bigcup_{i=1}^{N} (y\colon \alpha_i y - G(\alpha_i) > s)\right) \\
&\le \sum_{i=1}^{N} \int \exp(\lambda(h)(\alpha_i y - G(\alpha_i) - s))\,\mu^h(\mathrm{d}y) \\
&= \sum_{i=1}^{N} \exp(\lambda(h)(\lambda(h)^{-1} G^h(\lambda(h)\alpha_i) - G(\alpha_i)))\exp(-\lambda(h)s).
\end{aligned}$$

By virtue of (8.1) the estimate (8.2) follows. The theorem is proved.

A convex function L is called strictly convex in point β_0, if there exists such α_0 that

$$L(\beta) > L(\beta_0) + \alpha_0(\beta - \beta_0), \qquad \beta \neq \beta_0.$$

Theorem 8.2 (Gärtner). Let function L be strictly convex in point x. Then for any $\delta, \gamma > 0$ there exists such α_0, that for $h \leq h_0$ the inequality holds

$$\mu^h(y\colon\ \rho(y,x) < \delta) \geq \exp(-\lambda(h)(L(x) + \gamma)). \tag{8.3}$$

In problems that concern large deviations one usually has information about G rather than L. Thus, it is reasonable to express the strict convexity property of L in terms of function G.

Recall that function G is called sufficiently smooth if the set $(\alpha\colon\ G(\alpha) < \infty)$ has internal points, G is differentiable in all such points, and $\alpha_i \to \partial(\alpha\colon\ G(\alpha) < \infty)$, $i \to \infty$, $G(\alpha_i) < \infty$, implies $|\nabla G(\alpha_i)| \to \infty$; if $\partial(\alpha\colon\ G(\alpha) < \infty) = \varnothing$, then one may omit the last assumption $|\nabla G(\alpha_i)| \to \infty$.

Theorem 8.3 (Rockafellar (1970), Theorem 26.3). If function G is sufficiently smooth then function L is strictly convex in any point internal for the set $\{\beta\colon\ L(\beta) < \infty\}$ w.r.t. its linear hull.

Theorem 8.2 implies the following.

Theorem 8.4 (Gärtner). Let all assumptions about μ^h, G^h and G be satisfied, and function L is strictly convex in points dense in $(\beta\colon\ L(\beta) < \infty)$. Then for any $\delta, \gamma > 0$ and x (8.3) holds.

The couple of assertions (8.2) and (8.3) which hold for any $x \in E^l$ and $s > 0$ constitute the principle of large deviations in Freidlin–Wentzell form.

Theorems 8.4 and 8.3 imply the sufficient condition for (8.3): it is sufficient if function G is finite and differentiable everywhere.

We need some localization of Theorem 8.3. Suppose function L is not strictly convex in point β_0 and β_0 is internal for the set $(\beta\colon\ L(\beta) < \infty)$ w.r.t. the linear hull. Then there exists such α_0 that $G(\alpha_0) = \alpha_0\beta_0 - L(\beta_0)$ and $G(\alpha_0) = \alpha_0\beta - L(\beta)$. It means that function G is not differentiable in α_0. Indeed, if L is not strict convex then the graph of L contains the (non-vertical) segment to which the point $(\beta_0, L(\beta_0))$ belongs and this segment himself belongs to the hyperspace with the normal α_0. Thus, there exist the infinite number of support hyperspaces in $(\alpha_0, G(\alpha_0))$ for the graph of G, that is, G is not differentiable in α_0. In other words, the following theorem holds:

Theorem 8.5 (i). If β_0 is the internal point of the set $(\beta\colon\ L(\beta) < \infty)$ w.r.t. the linear hull and function L is not strictly convex in point β_0 then there exists such α_0 that $G(\alpha_0) = \alpha_0\beta_0 - L(\beta_0)$, and G is not differentiable in α_0.

In the other words, if G is differentiable in some point α, where $G(\alpha) = \alpha\beta_0 - L(\beta_0)$, then L is strictly convex in β_0.

Let us now consider the following situation. Let function G be convex, finite in the ball $(\alpha\colon |\alpha| \le R)$, differentiable in any point α, $|\alpha| < R$, and equal to $+\infty$ for any $|\alpha| \ge R$ (note that such function is lower semicontinuous). Let β_0 be an internal point for the set $(\beta\colon L(\beta) < \infty)$ w.r.t. its linear hull and $G(\alpha_0) = \alpha_0\beta_0 - L(\beta_0)$ for some $|\alpha_0| < R$. Then the following theorem holds true:

Theorem 8.5 (ii). Let the graph L not consist of a ray along any direction which contains zero. If function G is differentiable in the open ball $(\alpha\colon |\alpha| < R)$, is finite as $|\alpha| \le R$, and is equal to $+\infty$ as $|\alpha| > R$, then function L is strictly convex in β_0.

Theorems 8.5 (i) and 8.5 (ii) are already proved. Theorem 8.4 follows from Theorem 8.2.

Proof of Theorem 8.2. Function L is strictly convex in the point x. Let α_0 be such a point that $L(\beta) > L(x) + \alpha_0(\beta - x)$ as $\beta \ne x$. Then

$$G(\alpha_0) = \sup(\alpha_0\beta - L(\beta)) = \alpha_0 x - L(x).$$

Strict convexity of L in x implies inequality $G(\alpha_0) < \infty$. So $G^h(\lambda(h)\alpha_0) < \infty$ as $h \to 0$. Let us consider for such h the probabilistic measure

$$\mu^{h,\alpha_0}(\Gamma) = \int_\Gamma \exp(\lambda(h)(\alpha_0 y - G^h(\lambda(h)\alpha_0)))\,\mu^h(dy),$$

which is equivalent to the measure μ^h. We have

$$\mu^h(y\colon \rho(y,x) < \delta) = \int_{(y\colon \rho(y)<\delta)} \exp(-\lambda(h)(\alpha_0 y - G^h(\lambda(h)\alpha_0)))\,\mu^{h,\alpha_0}(dy).$$

Let $\delta' = \min(\delta, \gamma/3|\alpha_0|)$. Then

$$\begin{aligned}\mu^h(y\colon \rho(y,x) < \delta) &\ge \mu^{h,\alpha_0}(y\colon \rho(y,x) < \delta')\\ &\quad\times \exp(-\lambda(h)\{\alpha_0 x - \lambda(h)^{-1}G^h(\lambda(h)\alpha_0)\})\exp(-\lambda(h)\gamma/3)\\ &\ge \exp(-\lambda(h)(L(x) + 2\gamma/3))\,\mu^{h,\alpha_0}(y\colon \rho(y,x) < \delta').\end{aligned}$$

Let us check that $\mu^{h,\alpha_0}(y\colon \rho(y,x) < \delta') \to 1$ as $h \to 0$. Let us apply Theorem 8.1 to the family of measures (μ^{h,α_0}) for this aim. We have

$$\begin{aligned}G^{h,\alpha_0}(\alpha) &= \ln\int \exp(\alpha y)\,\mu^{h,\alpha_0}(dy) = G^h(\alpha + \lambda(h)\alpha_0) - G^h(\lambda(h)\alpha_0),\\ G^{\alpha_0}(\alpha) &= \lim_{h\to 0}\lambda(h)^{-1}G^h(\lambda(h)\alpha) = G(\alpha + \alpha_0) - G(\alpha_0),\\ L^{\alpha_0}(\alpha) &= L(\beta) - (\alpha_0\beta - G(\alpha_0)).\end{aligned}$$

Function $L^{\alpha_0}(\beta)$ is non-negative (as $G^{\alpha_0}(0) = 0$), and it equals to zero as $\beta = x$. Function $L^{\alpha_0}(\beta)$ is strictly convex in the point x as well as $L(\beta)$. Now, by choosing of α_0 and by virtue of representation of function L^{α_0} with the help of L we have for $\tilde{\beta} \neq x$

$$\begin{aligned} L^{\alpha_0}(\beta) - L^{\alpha_0}(x) &= L(\beta) - (\alpha_0\beta - G(\alpha_0)) - L(x) + (\alpha_0 x - G(\alpha_0)) \\ &= L(\beta) - L(x) - \alpha_0(\beta - x) > 0. \end{aligned}$$

Thus,

$$\gamma_0 = \min\left(L^{\alpha_0}(b) \colon \rho(\beta, x) \geq \delta'/2\right) > 0.$$

By virtue of Theorem 8.1 this implies for some $\gamma \in (0, \gamma_0)$, $s \in (\gamma, \gamma_0)$ and h sufficiently small

$$\mu^{h,\alpha_0}(y\colon \rho(y,x) \geq \delta') \leq \mu^{h,\alpha_0}(y\colon \rho(y, \Phi^{\alpha_0}(s)) \geq \delta'/2) \leq \exp\left(-\lambda(h)(s-\gamma)\right) \to 0.$$

This proves Theorem 8.2.

8.2. EXAMPLES

Example 8.1. How one can check the assertion (8.1)? The main and the simplest example for a Markov case (but not an independent one) is a finite Markov chain with a positive transition probabilities matrix (cf. Freidlin and Wentzell (1984), Chapter 7, Theorem 4.2, although this theorem deals with large deviations in averaging principle). Thus, (X_n) is a finite Markov chain with values in $Y = (y_1, \ldots, y_N)$, $y_i \in R^1$, $P = (p_{ij})_{i,j=1}^N$ is a transition probabilities matrix and $p_{ij} > 0$ for any i, j. We are interested in large deviation logarithmic asymptotics for sums $(X_1 + \ldots + X_n)$. The correct normalizing factor here is $\lambda(n) = \text{n}$. We have $G^n(\alpha) = \ln \mathsf{E}_x \exp(\alpha(X_1 + \ldots + X_n)/n)$, where $x = X_0$, so it is necessary to study whether the limit exists

$$G(\alpha) = \lim n^{-1} \ln \mathsf{E}_x \exp\left(\alpha(X_1 + \ldots + X_n)\right),$$

and whether this limit is differentiable in α. Let us consider the operator A on $B(Y)$:

$$Af(y_k) = \mathsf{E}_{y_k} f(X_1) \exp(\alpha X_1) = \sum_{i=1}^{\infty} f(y_i) \exp(\alpha y_i) p_{ki}.$$

Denote by $r(\alpha)$ the spectral radius of the matrix $P^\alpha = (p_{ki}^\alpha\colon\ p_{ki}^\alpha = \exp(\alpha) p_{ki})$. Operator A is positive, that is, if $f \geq 0$ then $Af \geq 0$. By virtue of Frobenius theorem for positive matrices $r(\alpha)$ is the eigen-value of A with positive eigen-function $e(x)$: $Ae(x) = r(\alpha) e(x)$. We have $0 < c = \min_x e(x) \leq \max_x e(x) = C < \infty$. Then

$$\begin{aligned} n^{-1} \ln \mathsf{E}_x \exp\left(\alpha(X_1 + \ldots + X_n)\right) &= n^{-1} \ln A^n 1(x) \leq n^{-1} \ln c^{-1} A^n e(x) \\ &= n^{-1} \ln c^{-1} r(\alpha)^n e(x) \\ &\leq n^{-1} \ln c^{-1} C + n^{-1} \ln r(\alpha)^n, \end{aligned}$$

and similarly

$$n^{-1}\ln \mathsf{E}_x \exp(\alpha(X_1 + \ldots + X_n)) \geq n^{-1}\ln C^{-1}c + n^{-1}\ln r(\alpha)^n.$$

These inequalities imply the existence of the limit $G(\alpha)$ and the validity of the representation $G(\alpha) = \ln r(\alpha)$. Function $\ln r(\alpha)$ is differentiable in α because the eigen-value $r(\alpha)$ is single and isolate, see Kato (1976).

Example 8.2. Another example includes the previous one: for any x, y the inequality

$$\mathsf{P}_x(X_1 \in \Gamma) \leq C\mathsf{P}_y(X_1 \in \Gamma)$$

holds. We will not give the proof now (see Stroock (1984), Chapter 6). It will follow from the proofs of more general results for Markovian and also stationary cases (see below). More general situation have been considered by Orey and Pelikan (1988). They generalized Stroock's result for the stationary case as well.

Example 8.3. Further example: let the process X_n satisfies the stochastic-difference equation

$$X_{n+1} = X_n + a(X_n) + b(X_n)W_n,$$

where (W_n) are i.i.d. random values with positive density, such that $\mathsf{E}W_n = 0$ and for any $\alpha > 0$

$$\mathsf{E}\exp(\alpha|W_n|) < \infty.$$

We assume that the matrix b is bounded and non-degenerate and the following stability condition holds true: for any $\alpha > 0$ there exist such a function $h > 0$ and a constant $\delta > 0$ that

$$\mathsf{E}_{X_0} \exp(\alpha\varphi(X_1))\, h(X_1) \leq (1-\delta)\, h(X_0).$$

(Note that often in examples $h(x) = \exp(\varepsilon|x|)$.) As K_m we choose the balls $\{x\colon |x| \leq m\}$, and $h_m(x) = h(x)$.

Let us explain how to check the necessary properties. Consider the process

$$z_n = \exp\left(\alpha\sum_{i=1}^{n}\varphi(X_i)\right) h(X_n).$$

By virtue of conditions on (W_n), a and b every z_n has a finite expectation. Let

$$\tau_m = \inf(n \geq 0\colon\ X_n \in K_m).$$

As $n < \tau_m$ we have $(\gamma > 0, m \geq m_\gamma)$

$$\begin{aligned}\mathsf{E}(z_{n+1}|F_n) &= \mathsf{E}\left(\exp(\alpha\sum_{i=1}^{n+1}\varphi(X_i))\, h(X_{n+1})|F_n\right)\\ &= \exp(\alpha S_n)\,\mathsf{E}(\exp(\alpha\varphi(X_{n+1}))\, h(X_{n+1})|F_n)\\ &\leq \exp(\alpha S_n)(1-\delta)\, h(X_n).\end{aligned}$$

Further calculations are similar to those in Chapter 5.5.

8.3. MARKOVIAN NON-COMPACT CASE

Let X_n, $n \geq 0$, be a homogeneous Markov process with values in E^d, $U \in \mathcal{B}(E^l)$, φ: $E^d \to E^l$ is a Borel function, $S_n := \sum_{k=0}^{n-1} \varphi(X_k)$, d, $l \geq 1$. The logarithmic asymptotics of probabilities $\mathsf{P}(S_n/n \in U)$, $n \to \infty$, is studied. Let us consider the limit

$$G(\alpha) := \lim_{n\to\infty} n^{-1} \ln \mathsf{E}_x \exp(\alpha S_n). \tag{8.4}$$

If this limit exists then the answer is formulated in terms of Fenchel–Legendre transformation of function G. Let us consider the semigroup of operators on $B(E^d)$:

$$(A^\alpha)^n f(y) := \mathsf{E}_y \exp(\alpha S_n) f(X_n).$$

Assume that the following stability and non-degeneracy conditions are fulfilled. There exist such a sequence $(K^m,\ m \to \infty)$ of bounded Borel sets $K^m \uparrow E^d$, that:

(A) for any $a > 0$, $m \geq 0$ there exist such locally bounded functions $h^+_{m,a}(x) \geq h^-_{m,a}(x) \geq 0$, $x \in E^d$, that $1/h^\pm$ are bounded on K^m and for any $a > 0$, $x \in E^d$

$$h^-_{m,a}(x) \leq \mathsf{E}_x \exp(a\tau_m) \leq h^+_{m,a}(x),$$

where $\tau_m := \inf(k \geq 1\colon\ X_k \in K^m)$;

(B) for any $\alpha > 0$ and $m \geq m_a$, $x \in K^m$, $n \geq 0$

$$C^{-1}_{m,a} \leq \mathsf{E}_{0,x} h^+_{m,a}(X_n) \leq C_{m,a} \qquad (C_{m,a} > 0), \text{ and } \lim_{m\to\infty} C_{m,a} = 1;$$

(A_φ) for any $a > 0$ there exist such locally bounded functions $h^+_{m,a}(x)$, $h^-_{m,a}(x)$, $x \in E^d$, that $1/h^\pm_{m,a}$ are locally bounded and for any $|\alpha| < a$, $m \geq m_a$, $x \in E^d$

$$h^-_{m,a}(x) \leq \mathsf{E}_{0,x} \exp\left(\alpha \sum_{k=1}^{\tau_m} \varphi(X_k)\right) \leq h^+_{m,a}(x);$$

(B_φ) for any $a > 0$, $m \geq m_a$, $x \in K^m$, $n \geq 0$

$$C^{-1}_{m,a} \leq \mathsf{E}_{0,x} h^-_{m,a}(X_n) \leq \mathsf{E}_{0,x} h^+_{m,a}(X_n) \leq C_{m,a} \text{ and } \lim_{m\to\infty} C_{m,a} = 1;$$

(C) $\liminf\limits_{m\to\infty,\ x\in K^m} \mathsf{P}_{0,x}(X_1 \in K^m) = 1$;

(D) for any $m \geq m_0$

$$\inf_{x,y\in K^m} \int_{K^m} 1 \wedge \frac{\mathsf{P}(y,1,dz)}{\mathsf{P}(x,1,dz)} \mathsf{P}(x,1,dz) > 0,$$

where $d\mathsf{P}'/d\mathsf{P}$ means the density of the absolute continuous component of measure P' with respect to P;

(E) the process $(X_{\tau^n_m}, n \geq 0)$ has a stationary distribution μ_m on K^m, for any $m \geq m_0$ which is ergodic in the following sense:

$$\sup_{x\in K^m} \operatorname{var}(\mathcal{L}(X_n|X_0 = x) - \mu_m) \to 0, \qquad n \to \infty,$$

where $\tau_m^1 = \tau_m$, $\tau_m^{i+1} = \theta_{\tau_m}\tau_m^i$, $i \geq 1$;

(*F*) for any $V \in \mathcal{B}(E^d)$ the probability $\mathsf{P}_y(X_1 \in V \cap K^m)$ is continuous in y for all m.

Condition (*F*) is satisfied, for example, for non-degenerate diffusion with bounded measurable coefficients (see, Krylov and Safonov (1981)). Condition (*D*) is a local Doeblin-type one.

Theorem 8.6. Let conditions (*A*), (*B*), (*C*), (*E*) and (*F*) be satisfied. Then for any bounded φ and any $\alpha \in E^l$ the limit (8.4) exists, function G does not depend on x and is convex, and for any $\delta > 0$, $a > 0$, $|\alpha| < a$, $m \geq m(\delta, a)$ and some $C(m,a) > 0$ and $C > 0$ for any $x \in E^d$ the following inequalities hold:

$$\begin{aligned}-\delta - C(m,a)n^{-1} &- C(m,a)\ln h_m^-(x) \\ &\leq G(\alpha) - n^{-1}\ln \mathsf{E}_x \exp(\alpha S_n) \\ &\leq \delta + C(m,a)n^{-1} + C(m,a)\ln h_m^+(x). \end{aligned} \tag{8.5}$$

Let us denote $L(z) := \sup(\alpha z - G(\alpha))$.

Theorem 8.7. Let conditions of the previous theorem be satisfied and φ is bounded. Then $nL(z)$ is an action functional (the rate function) for the family of random values (S_n/n), that is, for any $z \in E^l$, $\gamma > 0$, $\delta > 0$ there exists such $\varepsilon_0 > 0$, that for any $\varepsilon \in (0, \varepsilon_0)$ and $x \in E^d$

$$\mathsf{P}_x(|S_n/n - z| < \delta) \geq \exp(-n(L(z) + \gamma)), \tag{8.6}$$

and for any $s \geq 0$, $\gamma > 0$, $\delta > 0$, $\varepsilon \in (0, \varepsilon_0)$, $x \in E^d$

$$\mathsf{P}_x(\rho(S_n/n, \Phi(s)) > \delta) \leq \exp(-n(s - \gamma)), \tag{8.7}$$

where $\Phi(s) := (z \in E^l\colon L(z) \leq s)$ is a compact.

Theorem 8.8. Let conditions (A_φ), (*A*), (B_φ), (*B*), (*C*), (*D*), (*E*) and (*F*) be satisfied. Then for any $\alpha \in E^l$ the limit (8.4) exists, function G does not depend on x, is convex and inequalities (8.5) hold for the given φ.

Note that the differentiability of G is not asserted.

Theorem 8.9. Let conditions of Theorem 8.8 be satisfied. Then $nL(z)$ is the action functional for the family (S_n/n), that is, assertions (8.6) and (8.7) hold true.

8.4. AUXILIARY RESULTS

Set $\sigma_m = \sigma_m^1 := \inf(n \geq 1\colon X_{n-1} \in K^m, X_n \in K^m)$, $\sigma_{m,\mu} = \sigma_{m,\mu}^1 := \sigma_m^{1+\zeta}$, where ζ is a Poisson random value with the parameter $\mathsf{E}\zeta = \mu$ on some independent probability space, $\sigma_{m,\mu}^{n+1} := \vartheta_{\sigma_{m,\mu}}\sigma_{m,\mu}^n$, where $\sigma_{m,\mu}$ is built with the help of new independent Poisson value ζ.

Lemma 8.1. Let conditions (A_φ), (B_φ) and (*C*) be satisfied. Then for any $a > 0$, $|\alpha| < a$, $m \geq m_a$, x the following inequalities hold:

$$(h_{m,2a}^-(x))^{1/2}C^{-1} \leq \mathsf{E}_x \exp\left(\alpha \sum_{k=1}^{\sigma_m} \varphi(X_k)\right) \leq C(h_{m,2a}^+(x))^{1/2}.$$

Proof. Denote $\tau_m^0 := 0$, $\tau_m^1 := \tau_m$, $\tau_m^{n+1} := \vartheta_{\tau_m}\tau_m^n$, $n \geq 1$. Then

$$\begin{aligned}
\mathsf{E}_x \exp\left(\alpha \sum_{k=1}^{\sigma_m} \varphi(X_k)\right) &= \mathsf{E}_x \exp\left(\sum_{i=1}^{\infty} \sum_{k=\tau_m^i+1}^{\tau_m^{i+1}} \alpha\varphi(X_k)\right) \mathrm{I}(\tau_m^i < \sigma_m) \\
&\leq \sum_{i=1}^{\infty} \mathsf{E}_x \exp\left(\sum_{k=1}^{\tau_m^1} \alpha\varphi(X_k)\right. \\
&\quad \left. + \sum_{j=1}^{i} \sum_{k=\tau_m^i+1}^{\tau_m^{j+1}} \alpha\varphi(X_k)\right) \prod_{k=0}^{i} \mathrm{I}(X_{\tau_m^k+1} \notin K^m) =: \sum_{i=1}^{\infty} S(i).
\end{aligned}$$

By virtue of the Hölder inequality and the strong Markovian property we obtain

$$\begin{aligned}
S(i) &\leq \left[\mathsf{E}_x \exp\left(\sum_{k=1}^{\tau_m^1} 2\alpha\varphi(X_k) + \sum_{j=1}^{i} \sum_{k=1}^{\tau_m^{j+}} 2\alpha\varphi(X_k)\right)\right]^{1/2} \left[\mathsf{E}_x \prod_{k=0}^{i} \mathrm{I}(X_{\tau_m^k+1} \notin K^m)\right]^{1/2} \\
&\leq (1-p_m)^{(i+1)/2} \left[\mathsf{E}_x \exp\left(\sum_{k=1}^{\tau_m^1} 2\alpha\varphi(X_k) + \sum_{j=1}^{i} \sum_{k=1}^{\tau_m^{j+1}} 2\alpha\varphi(X_k)\right)\right]^{1/2} \\
&=: (1-p_m)^{(i+1)/2} S^1(i).
\end{aligned}$$

Again by virtue of the strong Markovian property and condition (A_φ) as $m \geq m_{2a}$ we obtain

$$\begin{aligned}
S^1(i)^2 &\leq \mathsf{E}_x \exp\left(\sum_{k=1}^{\tau_m^1} 2\alpha\varphi(X_k) + \sum_{j=1}^{i-1} \sum_{k=1}^{\tau_m^{j+1}} 2\alpha\varphi(X_k)\right) \mathsf{E}_{\tau_m^i, X_{\tau_m^i}} \exp\left(\sum_{q=1}^{\tau_m} 2\alpha\varphi(X_q)\right) \\
&\leq \mathsf{E}_x \exp\left(\sum_{k=1}^{\tau_m^1} 2\alpha\varphi(X_k) + \sum_{j=1}^{i-1} \sum_{k=1}^{\tau_m^{j+1}} 2\alpha\varphi(X_k)\right) h_{m,2a}^{+}(X_{\tau_m^j}) \\
&\leq C_{m,2a} \mathsf{E}_x \exp\left(\sum_{k=1}^{\tau_m^1} 2\alpha\varphi(X_k) + \sum_{j=1}^{i-1} \sum_{k=1}^{\tau_m^{j+1}} 2\alpha\varphi(X_k)\right).
\end{aligned}$$

By induction and due to (A_φ) and (B_φ) we conclude

$$S^1(i)^2 \leq C_{m,2a}^i \mathsf{E}_x \exp\left(\sum_{k=1}^{\tau_m^1} 2\alpha\varphi(X_k)\right) \leq C_{m,2a}^i h_{m,2a}^{+}(x).$$

Thus,

$$\mathsf{E}_x \exp\left(\alpha \sum_{k=1}^{\sigma_m} \varphi(X_k)\right) \leq \sum_{i=1}^{\infty} (1-p_m)^{(i+1)/2} C_{m,2a}^{i/2} [h_{m,2a}^{+}(x)]^{1/2}.$$

By virtue of (B_φ) one can choose such $m = m(2a)$ for a fixed than for any $m \geq m(2a)$

$$(1-p_m)C_{m,2a} \leq q, \qquad q < 1.$$

Then

$$\mathsf{E}_x \exp\left(\alpha \sum_{k=1}^{\sigma_m} \varphi(X_k)\right) \leq \frac{q}{1-q} [h_{m,2a}^{+}(x)]^{1/2}.$$

This implies the right-hand side of the desired inequality. Lemma 8.1 is proved.

Lemma 8.2. Let conditions of Lemma 8.1 be satisfied. Then for any $a > 0$, $|\alpha| < a$, $\mu_0 > 0$, $m \geq m_{a,\mu_0}$, $0 < \mu < \mu_0$ inequalities hold:

$$C^{-1}(h^{-}_{m,2(a+\mu_0)}(x))^{1/2} \leq \mathsf{E}_x \exp\left(\alpha \sum_{k=1}^{\sigma_{m,\mu}} \varphi(X_k)\right) \leq C(h^{+}_{m,2(a+\mu_0)}(x))^{1/2}.$$

Proof follows from the previous lemma.

Let $\alpha \in E^l$, $\mu > 0$, μ_m is the stationary distribution of $(X_{\tau^i_m})$ on K^m (see condition (E)). Let us denote $L^\alpha_m := (f(x), x \in K^m : f(x) = \exp(\alpha\varphi(x))g(x)$, $g \in B(K^m))$, $u_\alpha(x) = \exp(\alpha\varphi(x))$. Let us consider the family of operators $(A^\alpha_{m,\mu})$ on L^α_m:

$$A^\alpha_{m,\mu}f(x) = \mathsf{E}_x f(X_{\sigma_{m,\mu}}) \exp\left(\alpha \sum_{k=0}^{\sigma_{m,\mu}-1} \varphi(X_k)\right). \tag{8.8}$$

Denote by $\|\cdot\|_m$ the norm in L^α_m:

$$\|f\|_m := \sup_{x \in K^m} |f(x)\exp(-\alpha\varphi(x))|.$$

Lemma 8.3. Let conditions (A_φ) and (B_φ) be satisfied. Then for any $|\alpha| < a$, $m \geq m_{a,\mu}$ the operator $A^\alpha_{m,\mu}$ on L^α_m is upper u_α-bounded, that is (see Krasnoselskii, Lifshitz and Sobolev (1989)) for any $f \in L^\alpha_m$, $f \geq 0$, $f \not\equiv 0$ there exists such $c(f) > 0$ that for any $x \in K^m$

$$A^\alpha_{m,\mu}f(x) \leq c(f)\, u_\alpha(x).$$

Proof follows from Lemma 8.2 and conditions (A_φ) and (B_φ).

Lemma 8.4. Let conditions (A_φ), (B_φ), (D) and (E) be satisfied. Then for any $\mu > 0$, $|\alpha| < a$, $m \geq m_{a,\mu}$ the operator $A^\alpha_{m,\mu}$ on L^α_m is lower u_α-bounded, that is (see Krasnoselskii, Lifshitz and Sobolev (1989)) for any $f \in L^\alpha_m$, $f \geq 0$, $f \not\equiv 0$ there exists such $c'(f) > 0$ that for any $x \in K^m$ the inequality $c'(f)\, u_\alpha(x) \leq A^\alpha_{m,\mu}f(x)$ holds.

Proof follows from conditions and Lemma 8.2 by virtue of the inequality

$$\inf_{x \in K^m} \mathsf{P}_x(X_{\sigma_{m,\mu}} \in U) > 0$$

(see (F) and (D)).

Lemma 8.5. Let conditions (A_φ), (B_φ), (C) and (F) be satisfied. Then for any $|\alpha| < a$, $m \geq m_{a,\mu}$ the operator $A^\alpha_{m,\mu}$ on L^α_m is a monotonous-compact one (see Krasnoselskii, Lifshitz and Sobolev (1989)), that is, $A^\alpha_m \geq 0$ and if $f_1 \geq f_2 \geq \ldots \geq f_0$ and $f_i \in L^\alpha_m$, $i \geq 0$ then the sequence $(A^\alpha_m f_n)_{n \geq 1}$ converges in L^α_m-norm.

Proof follows from Lemma 8.2, Lebesgue convergence theorem and condition (F).

Lemma 8.6. Let conditions (A_φ), (B_φ), (C), (D), (E) and (F) be satisfied. Then for any $|\alpha| < a$, $\mu \leq \mu_0$, $m \geq m(a, \mu_0)$ the spectral radius $r(A^\alpha_{m,\mu})$ is the single isolated eigen-value of $A^\alpha_{m,\mu}$ on L^α_m and correspondent eigen-function e_m is positive (that is, $e_m \geq 0$, $e_m \in L^\alpha_m$, $e_m \not\equiv 0$). Function $G_{m,a}(\alpha) := \ln r(A^\alpha_{m,\mu})$ is differentiable in $\alpha(|\alpha| < a)$ and convex in $\alpha(\alpha \in E^l)$.

Proof. By virtue of Lemmas 8.2–8.4 and Theorem 9.7 (Krasnoselskii, Lifshitz and Sobolev (1989)) the operator $A^\alpha_{m,\mu}$ has the eigen-function $e^\alpha_m = e^\alpha_{m,\mu} \in L^\alpha_m, e_{\alpha,m} \geq 0$ with eigen-value $r(A^\alpha_{m,\mu})$. By virtue of Theorem 11.1(b) from Krasnoselskii, Lifshitz, Sobolev (1989) the eigen-value $r(A^\alpha_{m,\mu})$ is single. Then the differentiability in α follows from Kato (1976). The convexity follows from the properties of the functions exp and ln. Lemma 8.6 is proved.

Let us denote $l(n) := \sup(k\colon\ k = 0, 1, \ldots, \sigma^k_{m,\mu} \leq n)$.

Lemma 8.7. Let conditions (A_φ), (B_φ), (C), (D), (E) and (F) be satisfied. Then for any $m \geq m_0$, $\mu \leq \mu_0$

$$\mathsf{P}(l(n)/n \to \text{const} =: \nu_{m,\mu},\ n \to \infty) = 1,$$

where $\nu^{-1}_{m,\mu}$ equals to the expectation of the difference $\sigma^{n+1}_{m,\mu} - \sigma^n_{m,\mu}$ w.r.t. stationary measure for $(X_{\sigma^n_{m,\mu}})$.

Proof follows from the ergodic Birkhoff–Khinchin theorem and condition (E).

Lemma 8.8. Let conditions (A_φ), (B_φ), (C), (D), (E) and (F) be satisfied. Then for any α, $|\alpha| < a$, $m \geq m_a$, $\mu \leq \mu_a$ there exists the limit

$$\lim_{n\to\infty} n^{-1} \ln \mathsf{E}_x \exp(\alpha S_{\sigma^n_{m,\mu}}) =: G_{m,\mu}(\alpha),$$

function $G_{m,\mu}(\alpha)$ does not depend on x and for any $a > 0$, $\delta > 0$ there exist such $m(a, \delta)$, $\mu(a, \delta)$ that for any $m \geq m(a, \delta)$, $\mu \leq \mu(a, \delta)$, $|\alpha| < a$, n

$$|G_{m,\mu}(\alpha) - n^{-1} \ln \mathsf{E}_x \exp(\alpha S_{\sigma^n_{m,\mu}})| \leq \delta + C(m, \mu, a) n^{-1} |\alpha\varphi(x)|.$$

Proof. By virtue of Lemma 8.6 for m and μ^{-1} large enough the value $r(A^\alpha_{m,\mu})$ is the eigen-value and its eigen-function e^α_m on L^α_m is non-negative. By virtue of Lemma 8.2, condition (A_φ) and Lemma 9.1 from Krasnoselskii, Lifshitz and Sobolev (1989) it follows

$$\nu_a := \inf_{|\alpha|\leq a} r(A^\alpha_{m,\mu}) > 0.$$

By virtue of Lemmas 8.3 and 8.4

$$c_a u_a(x) \leq e_m(x) = r(A^\alpha_{m,\mu})^{-1} A^\alpha_{m,\mu} e_m(x) \leq C_a u_a(x).$$

Hence, $e^1_m(x) = u_\alpha(x)^{-1} e_m(x) \geq c_a > 0$ and $e^1_m(x) \leq C_a < \infty$. Then

$$C_a^{-1} r(A^\alpha_{m,\mu}) e^\alpha_m(x) \leq \mathsf{E}_x \exp(\alpha S_{\sigma_{m,\mu}}) \leq c_a^{-1} r(A^\alpha_{m,\mu}) e^\alpha_m(x),$$

thus, by induction we obtain for any n

$$C_a^{-1} r(A_{m,\mu}^{\alpha})^n e_m^{\alpha}(x) \le \mathsf{E}_x \exp(\alpha S_{\sigma_{m,\mu}^n}) \le c_a^{-1} r(A_{m,\mu}^{\alpha})^n e_m^{\alpha}(x).$$

Now the desired assertion follows from the Lemma 8.8.

Lemma 8.9. Let conditions (A), (B), (C), (D), (E) and (F) be satisfied. Then for any $\delta, C > 0$

$$\mathsf{P}\left(\left|\frac{l(n)}{n} - \nu_{m,\mu}\right| > \delta\right) \le \exp(-Cn),$$

if m and μ^{-1} are large enough and $n \ge n(\delta, x)$.

Proof of Lemma 8.9 follows from the estimates of Lemma 8.2.

8.5. PROOFS OF THEOREMS 8.6–8.8

Theorems 8.6 and 8.7 follow from Theorems 8.8 and 8.9 correspondently.

Proof of Theorem 8.8. Let us denote $T_{m,\mu}^n = \sigma_{m,\mu}^{[\nu_{m,\mu} n]}$. We have

$$\begin{aligned}
\mathsf{E}_x \exp(\alpha S_n) &= \mathsf{E}_x \exp\left(\alpha \sum_{k=0}^{n-1} \varphi(X_k)\right) \\
&= \mathsf{E}_x \exp(\alpha S_{T_{m,\mu}^n}) \exp(\alpha(S_n - S_{T_{m,\mu}^n})) \\
&\le \mathsf{E}_x \exp(\alpha S_{T_{m,\mu}^n}) \exp(a\|\varphi\||T_{m,\mu}^n - t|).
\end{aligned}$$

Let $\delta > 0$, $C > 0$. By virtue of Lemma 8.9 for m, μ^{-1} and n large enough

$$\mathsf{P}(|n - T_{m,\mu}^n| > \delta n) \le 2\exp(-Cn).$$

Thus,

$$\begin{aligned}
&\mathsf{E}_x \exp(\alpha\, S_{T_{m,\mu}^n}) \exp(a\|\varphi\||T_{m,\mu}^n - n|) \\
&\quad= \mathsf{E}_x \exp(\alpha S_{T_{m,\mu}^n}) \exp(a\|\varphi\||T_{m,\mu}^n - n|)\, \mathrm{I}(|n - T_{m,\mu}^n| > \delta n) \\
&\qquad + \mathsf{E}_x \exp(\alpha S_{T_{m,\mu}^n}) \exp(a\|\varphi\||T_{m,\mu}^n - n|)\, \mathrm{I}(|n - T_{m,\mu}^n| \le \delta n) \\
&\quad\le \exp(a\|\varphi\|n)\, \mathsf{P}(|n - T_{m,\mu}^n| > \delta n) \\
&\qquad + \mathsf{E}_x \exp(\alpha S_{T_{m,\mu}^n}) \exp(a\|\varphi\||T_{m,\mu}^n - n|)\, \mathrm{I}(|n - T_{m,\mu}^n| \le \delta n) \\
&\quad\le 2\exp(a\|\varphi\|n)\exp(-Cn) + \mathsf{E}_x \exp(\alpha S_{T_{m,\mu}^n}) \exp(a\|\varphi\|\delta n).
\end{aligned}$$

By virtue of Lemma 8.8

$$\limsup_{n\to\infty} n^{-1} \ln \mathsf{E}_x \exp(\alpha S_n) \le G_{m,\mu}(\alpha) + \delta.$$

Hence,

$$\limsup_{n\to\infty} n^{-1} \ln \mathsf{E}_x \exp(\alpha S_n) \le \liminf_{m\to\infty} \liminf_{\mu\to 0} G_{m,\mu}(\alpha).$$

Similarly we obtain the lower estimate

$$\liminf_{n\to\infty} n^{-1}\ln \mathsf{E}_x \exp(\alpha S_n) \geq \limsup_{m\to\infty} \limsup_{\mu\to 0} G_{m,\mu}(\alpha).$$

Thus, the limit exists

$$\lim_{n\to\infty} n^{-1}\ln \mathsf{E}_x \exp(\alpha S_n) = \lim_{m\to\infty} \limsup_{\mu\to 0} G_{m,\mu}(\alpha) = \lim_{m\to\infty} \liminf_{\mu\to 0} G_{m,\mu}(\alpha).$$

Function G is automatically convex in α. Desired estimates follow now from Lemma 8.8. Theorem is proved.

8.6. PROOF OF THEOREM 8.9

Function G is finite and convex, hence, it is continuous (see Rockafellar (1970)). Thus, the upper bound (8.7) follows from Theorem 8.8 and Gärtner 's theorem (Gärtner (1977), see also Theorem 5.1.1 in Freidlin and Wentzell (1984)).

Proof of the lower bound (8.6) follows the considerations in the proof in Gärtner (1977) or Freidlin and Wentzell (1984), Theorem 5.1.2. One must only keep in mind that the differentiability of G is not stated. Let us fix $z \in E^l$, $\delta > 0$, $\gamma > 0$. Function $L(z) = \sup(\alpha z - G(\alpha))$ is lower semicontinuous. (8.6) is trivial if $L(z) = +\infty$, so we may assume $L^G(z) < \infty$. Let us divide the proof into steps.

(A) Denote $A = (z'\colon\ L(z') < \infty)$, A_0 is the set of all internal elements from A with respect to the linear hull (see Rockafellar (1970)). Then there exists such $z_0 \in A_0$ that

$$|z - z_0| < \delta/4, \tag{8.9}$$

and

$$|L(z) - L(z_0)| < \gamma/4. \tag{8.10}$$

If $z_0 \in A_0$ then there exists such $\alpha_0 \in E^l$ that $L(z_0) = \alpha_0 z_0 - G(\alpha_0)$ and $z_0 \in \partial_\alpha G(\alpha_0)$. Let us consider the cone N generated by external normals to the epigraph *epi G* in the point α_0 (see Rockafellar (1976)). Let π be its affine hull in $R \times E^l$. Then N is a body cone in π, that is, its interior in π is not empty. Let us denote it by N_0. π is an affine hyperspace, $0 \leq \dim\pi \leq l+1$. We may choose such $z_0 \in \partial_\alpha G(\alpha_0)$ that $(-1, z_0) \in N_0$ and inequalities (8.9) and (8.10) are satisfied. Then $\sup(\alpha z_0 - G(\alpha)) = \alpha_0 z_0 - G(\alpha_0)$ as well. Let us denote $\pi^+ = \pi \cap E^l$, $\pi^- = (\pi^+)^\perp$ in E^l, $0 \leq k = \dim\pi^+ \leq 1$. Let us choose some orthonormed basis $(e_1, \dots, e_k)$ in π^+ and let $(e_{k+1}, \dots, e_l)$ be an orthonormed basis in π^-. Then function G in α_0 is differentiable in $(e_{k+1}, \dots, e_l)$ for π^+ is maximal, and $\partial_{(e_1,\dots,e_k)} G(\alpha_0) = \partial_\alpha G(\alpha_0) \cap \pi^+$ is a body set in π^+. We do not change designations of functions G and $G_{m,\mu}$ in new coordinates. Note that functions $G_{m,\mu}$ are defined maybe not for all α. We suppose that $|\alpha| < a, m$ and $\mu^{-1}$are large enough such that $G_{m,\mu}(\alpha)$ are well defined for $|\alpha| \leq a + 1$. Let $G_{m,\mu}(\alpha) := +\infty$ when $|\alpha| > a + 1$.

Let us build point $\alpha_{m,\mu}$ and vector $z_{m,\mu}$ in the following way.

(1) $(\alpha^{k+1}_{m,\mu}, \dots, \alpha^l_{m,\mu}) := (\alpha^{k+1}_0, \dots, \alpha^l_0)$,

(2) $(z^1_{m,\mu}, \dots, z^k_{m,\mu}) := (z^1_0, \dots, z^k_0)$,

(3) $(\hat{\alpha}^1_{m,\mu}, \ldots, \hat{\alpha}^k_{m,\mu}) := \arg \min_{(\alpha^1, \ldots, \alpha^k)} ((\alpha^1, \ldots, \alpha^k)(z^1, \ldots, z^k) - G_{m,\mu}(\alpha^1, \ldots, \alpha^k, \alpha_0^{k+1}, \ldots, \alpha_0^l))$,
$\hat{\alpha}_{m,\mu} := (\alpha^1_{m,\mu}, \ldots, \alpha^k_{m,\mu}, \alpha_0^{k+1}, \ldots, \alpha_0^l)$,

(4) $\alpha_{m,\mu} := \hat{\alpha}_{m,\mu} \mathrm{I}(|\hat{\alpha}_{m,\mu} - \alpha_0| < 1) + \alpha_0 \mathrm{I}(|\hat{\alpha}_{m,\mu} - \alpha_0| \geq 1)$,

(5) $(z^{k+1}_{m,\mu}, \ldots, z^l_{m,\mu}) := \nabla_{(\alpha^{k+1}, \ldots, \alpha^l)} G_{m,\mu}(\alpha_{m,})$,
$z_{m,\mu} := (z^1_{m,\mu}, \ldots, z^l_{m,\mu})$.

We have $G_{m,\mu}(\alpha) \to G(\alpha)$ locally uniformly if $m, \mu^{-1} \to \infty$. Let us show that $\alpha_{m,\mu} \to \alpha_0$ and $z_{m,\mu} \to z_0$ when $m, \mu^{-1} \to \infty$. By virtue of (1) and (2) we have $(\alpha^{k+1}_{m,\mu}, \ldots, \alpha^l_{m,\mu}) = (\alpha_0^{k+1}, \ldots, \alpha_0^l)$, $(z^1_{m,\mu}, \ldots, z^k_{m,\mu}) = (z_0^1, \ldots, z_0^k)$. Function $G(\cdot, \alpha_0^{k+1}, \ldots, \alpha_0^l)$ (as well as $G_{m,\mu}$) is differentiable in $(\alpha^1, \ldots, \alpha^k)$ in the point $(\alpha_0^{k+1}, \ldots, \alpha_0^l)$. Let the generalized sequence $((\alpha^1_{m,\mu}, \ldots, \alpha^k_{m,\mu}), m, \mu^{-1} \to \infty)$ have a limit point $(\bar{\alpha}^1, \ldots, \bar{\alpha}^k) \neq (\alpha_0^1, \ldots, \alpha_0^k)$. Then for vector $(z_0^1, \ldots, z_0^k)$ belongs to the interior of the subgradient $\partial_{(\alpha^1, \ldots, \alpha^k)} G(\alpha_0)$ in the correspondent hyperspace E^k and G is convex then

$$(z_0^1, \ldots, z_0^k) \notin \partial_{(\alpha^1, \ldots, \alpha^k)} G(\bar{\alpha}^1, \ldots, \bar{\alpha}^k, \alpha_0^{k+1}, \ldots, \alpha_0^l),$$

but on the other hand

$$(z_0^1, \ldots, z_0^k) = \nabla_{(\alpha^1, \ldots, \alpha^k)} G_{m,\mu}(\alpha^1_{m,\mu}, \ldots, \alpha^k_{m,\mu}, \alpha_0^{k+1}, \ldots, \alpha_0^l),$$

and by virtue of subgradients' convergence (see Rockafellar (1970)) we obtain

$$(z_0^1, \ldots, z_0^k) \in \partial_{(\alpha^1, \ldots, \alpha^k)} G(\bar{\alpha}^1, \ldots, \bar{\alpha}^k, \alpha_0^{k+1}, \ldots, \alpha_0^l).$$

This contradiction shows that

$$\lim_{m,\mu^{-1} \to \infty} (\alpha^1_{m,\mu}, \ldots, \alpha^k_{m,\mu}) = (\alpha_0^1, \ldots, \alpha_0^k).$$

Thus,

$$\alpha_{m,\mu} \to \alpha_0, \qquad m, \mu^{-1} \to \infty.$$

Recall that $|\alpha_{m,\mu} - \alpha_0| < 1$. Denote $L_{m,\mu}(z) := \sup_\alpha (\alpha z - G_{m,\mu}(\alpha))$. Then

$$\begin{aligned} |L(z_0) - L_{m,\mu}(z_{m,\mu})| &= |L(z_0) - \sup_\alpha (\alpha z_{m,\mu} - G_{m,\mu}(\alpha)| \\ &= |L(z_0) - (\alpha_{m,\mu} z_{m,\mu} - G_{m,\mu}(\alpha_{m,\mu})|. \end{aligned}$$

So by virtue of locally uniform convergence $G_{m,\mu} \to G$ we obtain

$$\lim_{m,\mu^{-1} \to \infty} |L(z_0) - L_{m,\mu}(z_{m,\mu})| = 0.$$

Thus, the right-hand side of (8.6) has the upper bound

$$\exp(-\varepsilon^{-1}(L(z) + \gamma)) \leq \exp(-\varepsilon^{-1}(L_{m,\mu}(z_{m,\mu}) + \gamma/2)) \tag{8.11}$$

if m and μ^{-1} are large enough and $|z - z_{m,\mu}| \leq \delta/3$.

(B) Let $m \geq 1$, $\mu > 0$. Consider the process $S_{\sigma^n_{m,\mu}}$, $n \geq 1$. It follows from the considerations in the proof of Theorem 8.8 that for any $c > 0$ and $\delta > 0$ and sufficiently large m and μ^{-1} the following inequality holds:

$$\mathsf{P}(n^{-1}|S_n - S_{\sigma^n_{m,\mu}}| > \delta/3) \leq \exp(-cn). \tag{8.12}$$

Then

$$\begin{aligned}\mathsf{P}(|n^{-1}S_n - z| < \delta) &\geq \mathsf{P}(n^{-1}|S_n - S_{\sigma^n_{m,\mu}}| > \delta/3) - \mathsf{P}(|n^{-1}S_{\sigma^n_{m,\mu}} - z_{m,\mu}| > \delta/3) \\ &\geq \mathsf{P}(|n^{-1}S_{\sigma^n_{m,\mu}} - z_{m,\mu}| > \delta/3) - \exp(-cn).\end{aligned}$$

(C) We choose $a > |\alpha_0| + 1$, $m \geq m(a, \delta/3)$, $\mu \leq \mu(a, \delta/3)$ such that function $G_{m,\mu}$ is differentiable in any $|\alpha| < a + 1$. If $L_{m,\mu}$ is not strictly convex in $z_{m,\mu}$ then $G_{m,\mu}$ cannot be differentiable in $\alpha_{m,\mu}$ (cf. Theorem 8.5). So $L_{m,\mu}$ is strictly convex in point $z_{m,\mu}$, that is,

$$L_{m,\mu}(z') > L_{m,\mu}(z_{m,\mu}) + \alpha_{m,\mu}(z' - z_{m,\mu}), \qquad z' \neq z_{m,\mu}.$$

Hence,

$$G_{m,\mu}(\alpha_{m,\mu}) = \alpha_{m,\mu}z_{m,\mu} - L_{m,\mu}(z_{m,\mu}) = \sup_{\alpha}(\alpha z_{m,\mu} - L_{m,\mu}(\alpha)). \tag{8.13}$$

Let

$$G^n_{m,\mu}(\alpha) := \ln E\exp(\alpha n^{-1}S_{\sigma^n_{m,\mu}})$$

and consider the probabilistic measure $\tilde{\mu}^n_{m,\mu}$:

$$\tilde{\mu}^n_{m,\mu}(\Gamma) := \mathsf{E}\exp(n\alpha_{m,\mu}S_{\sigma^n_{m,\mu}}/n - G^n_{m,\mu}(n\alpha_{m,\mu}))\,\mathrm{I}(n^{-1}S_{\sigma^n_{m,\mu}} \in \Gamma)$$

(see Freidlin and Wentzell (1984), Section 5.1). Then we obtain

$$\begin{aligned}&\mathsf{P}(|n^{-1}S_{\sigma^n_{m,\mu}} - z_{m,\mu}| < \delta/3) \\ &\quad= \mathsf{E}\,\mathrm{I}(|n^{-1}S_{\sigma^n_{m,\mu}} - z_{m,\mu}| < \delta/3) \\ &\quad= \int\limits_{y:|y-z_{m,\mu}|<\delta/3} \exp(-n\alpha_{m,\mu}y + G_{m,\mu}(n\alpha_{m,\mu}))\,\tilde{\mu}^n_{m,\mu}(dy).\end{aligned}$$

Analogous to Freidlin and Wentzell (1984), Section 5.1, let

$$\delta' := (\delta/3) \wedge (\gamma/(6|\alpha_{m,\mu}|)).$$

Then we have

$$\begin{aligned}&\mathsf{P}(|n^{-1}S_{\sigma^n_{m,\mu}} - z_{m,\mu}| < \delta/3) \\ &\quad\geq \tilde{\mu}^n_{m,\mu}(y\colon |y - z_{m,\mu} - z_{m,\mu}| < \delta') \\ &\quad\quad\times \exp(-n(\alpha_{m,\mu}z_{m,\mu} - n^{-1}G^n_{m,\mu}(n\alpha_{m,\mu})))\exp(-n\gamma/6).\end{aligned}$$

By virtue of (8.13) and Lemma 8.8 for n sufficiently large we find

$$\exp(-n(\alpha_{m,\mu}z_{m,\mu} - n^{-1}G^n_{m,\mu}(n\alpha_{m,\mu}))) \geq \exp(-n(L_{m,\mu}(z_{m,\mu}) + \gamma/3)).$$

So it remains to prove that

$$\tilde{\mu}^n_{m,\mu}(y\colon\ |y - z_{m,\mu}| < \delta') \to 1, \qquad n \to \infty. \tag{8.14}$$

By virtue of strict convexity of $L_{m,\mu}$ in $z_{m,\mu}$ (8.14) follows from considerations in the proof of Theorem 5.1.2 in Freidlin and Wentzell (1984) which use Theorem 5.1.1 from Freidlin and Wentzell (1984) about the upper estimate for the family of measures $(\tilde{\mu}^n_{m,\mu},\ n \geq 1)$. Theorem 8.9 is proved.

Remark 8.1. From assumptions of Theorem 8.9 and arguments from Ellis (1985) it follows that $L(z) = 0$ in the unique point $z_0 = \mathsf{E}^{\text{inv}} X_1$ (the expectation w.r.t. the invariant measure). Indeed, let $\delta > 0$ and $\varepsilon > 0$. Let $z_0 = \mathsf{E}^{\text{inv}} X_1 = 0$ for simplicity. Then $G(\alpha) \geq 0$ for any α. For any $c > 0$ as we know there exist such $m = m(\delta, c)$, $\mu = \mu(\delta, c)$ that for any $n > 0$

$$\mathsf{P}(|S_n - S_{\sigma^n_{m,\mu}}| > \delta n) \leq \exp(-cn).$$

Fix some $c > \|f\|$. Then

$$\begin{aligned} |G_n(\varepsilon)| &= |n^{-1} \ln \mathsf{E} \exp(\varepsilon S_n)| \\ &\leq n^{-1} |\ln \mathsf{E} \exp(\varepsilon S_{\sigma^n_{m,\mu}} + \varepsilon\delta n) + \exp(\varepsilon \|f\| n)\, \mathrm{I}\,(|S_n - S_{\sigma^n_{m,\mu}}| > \delta n)| \\ &\leq 2 \max(\varepsilon\delta + n^{-1} |\ln \mathsf{E} \exp(\varepsilon S_{\sigma^n_{m,\mu}})|; |\varepsilon\|f\| - c|). \end{aligned}$$

Thus,

$$G(\varepsilon) \leq 2\max(\varepsilon\delta + G_{m,\mu}(\varepsilon); \varepsilon\|f\| - c) = 2\varepsilon\delta + 2G_{m,\mu}(\varepsilon).$$

Now, $G_{m,\mu}$ is differentiable, and, moreover, $G'_{m,\mu}(0) = \mathsf{E}^{\text{inv}} S_{\sigma_{m,\mu}}$ tends to zero as $m \to \infty$, $\mu \to 0$. So we may suppose that $|G'_{m,\mu}(0)| < \delta$. Since $G'_{m,\mu}(0) = 0$, we obtain

$$G(\varepsilon) \leq 4\delta\varepsilon, \qquad \varepsilon \to 0,$$

Similarly,

$$G(\varepsilon) \geq -4\delta\varepsilon, \qquad \varepsilon \to 0.$$

Thus, the derivative $G'(0)$ does exist and $G'(0) = 0$.

Comments

Results of this chapter are similar to some results of Donsker and Varadhan (1975a, 1975b, 1977, 1983). Nevertheless they are obtained by another method, namely, Freidlin's method. This will play the role in the last chapter. Other similar results are in Ney and Nummelin (1987), these results are established with the help of the Donsker–Varadhan method and the conditions are similar but not completely equal. Large deviations for 'uniform' Doeblin-type conditions were studied be Iscoe, Ney and Nummelin (1985). Our form of conditions in terms of stability and recurrency seems to be convenient for both stochastic recurrent and stochastic differential equations, see Veretennikov (1990) and (1991b). A similar collection of lemmas are used for the 'compact' case by Kolmogorov (1949) in another problem.

In connection with the stochastic approximation we note that in Freidlin's method the differentiability of function G in α is usually assumed. This did not allow previous studies on large deviations for standard stochastic approximation algorithms with averaging in non-compact state space: see Dupuis and Kushner (1989) about this problem. Our result does not use the differentiability and it is important in the last chapter. We do not know whether the differentiability always holds.

Note also that our method works well in the non-Markovian stationary case as well as Donsker and Varadhan's one – see the next chapter. The exposition is based on papers by Veretennikov (1990) and (1991b) (in which diffusions are considered). Some versions of Theorems 8.6 and 8.7 were announced in Veretennikov (1989a), which is the result of a talk at the School-seminar on random processes theory (Preila, Lithuania, (1987)).

When the book was being prepared the paper by de Acosta, Ney and Nummelin (1991) became known to the authors. This paper also contains the theorem about a lower bound in large deviation principle for measure sequences without a differentiability assumption and the application to countable Markov chains. In this paper another approximation method is applied based on regularization with the help of a small additional Gaussian component.

Chapter 9

Large deviations for stationary processes

9.1. COMPACT NON-SINGULAR CASE

Consider the regular stationary process $(X_n, -\infty < n < \infty)$ with values in R^d(see Ibragimov and Linnik (1971)) and partial sums $S_n = \sum_{k=1}^{n} \varphi(X_k)$, where φ is a bounded measurable function, φ: $R^d \to R^l$.

Let μ be a stationary measure for X_n. By virtue of Birkhoff–Khinchin ergodic theorem $n^{-1}S_n \to \mathsf{E}^{\mu}\varphi(X_1)$. As earlier the logarithmic asymptotics of large deviations may be investigated if we state the limit

$$G(\alpha) = \lim_{n\to\infty} n^{-1} \ln \mathsf{E}_x \exp(\alpha S_n) \tag{9.1}$$

for any $\alpha \in E^l$, where x in E_x means all past paths of X, that is, $x = (X_0, X_{-1}, \ldots)$, and $G(\alpha)$ does not depend on x. The existence of the limit (9.1) in the Markovian case may be established with the help of Frobenius-type theorem (see previous chapter). In this chapter another method is used. This method is similar to one used by Stroock (1984), Chapter 6, although Stroock considers the Markovian case only and does not study the limit (9.1). The main assumption in Stroock (1984), Chapter 6, is the inequality

$$\mathsf{P}(X_1 \in \Gamma | X_0 = x) \le C\mathsf{P}(X_1 \in \Gamma | X_0 = y) \tag{9.2}$$

for some $C > 0$ and any $x, y \in E^d$, $\Gamma \in \mathcal{B}^d$. The following condition is equivalent to (9.2) in the Markovian case and is very useful in the non-Markovian case too:

$$\mathsf{P}(X_1^{\infty} \in \Gamma | X_{-\infty}^{0} = x) \le C\mathsf{P}(X_1^{\infty} \in \Gamma | X_{-\infty}^{0} = y) \tag{9.3}$$

for any $x, y \in (E^d)^{Z_-}$, $\Gamma \in \mathcal{B}((E^d)^{Z_+})$, $Z_- = (0, -1, -2, \ldots)$, $Z_+ = (1, 2, \ldots)$ with some $C > 0$. Let us introduce another condition which is also equivalent to (9.2) in the Markovian case but is not as restrictive in the non-Markovian case as (9.3). Assume that for any $x, y \in (E^d)^{Z_-}$, $\Gamma_N \in \mathcal{B}((E^d)^N)$ the inequality holds

$$\mathsf{P}(X_1^N \in \Gamma_N | X_{-\infty}^{0} = x) \le C_N \mathsf{P}(X_1^N \in \Gamma_N | X_{-\infty}^{0} = y), \tag{9.4}$$

where $C_N > 0$, $N = 1, 2, \ldots$, and

$$\lim_{N\to\infty} N^{-1} \ln C_N = 0. \tag{9.5}$$

Later we consider some examples when (9.4)–(9.5) and (9.3) are satisfied.

We study here bounded functions φ only. A more general case will be considered in the next section. Note that (9.4)–(9.5) is slightly stronger than the condition

$(R\text{–}M)$ in Orey and Pelikan (1988) but their considerations are based on another method.

Theorem 9.1. Let conditions (9.4)–(9.5) be satisfied and function φ is bounded. Then for any $\alpha \in E^l$ there exists the limit (9.1) uniformly in x, this limit does not depend on x and function G is finite everywhere and convex.

Denote by $L(z)$, $z \in E^l$, the Fenchel–Legendre transformation of $G(\alpha)$: $L(z) = \sup(\alpha z - G(\alpha))$. Function L is non-negative because $G(0) = 0$, $L(0) = 0$, and L takes its values from $[0, +\infty]$.

Theorem 9.2. Let conditions of Theorem 9.1 be satisfied. Then $nL(z)$ is the action functional (or rate function) for the family $(n^{-1}S_n)$, that is, for any $z \in E^l$, $\delta, \gamma > 0$ and $s > 0$ the following inequalities hold:

$$\mathsf{P}(|n^{-1}S_n - z| < \delta) \geq \exp(-n(L(z) - \gamma)), \qquad n \to \infty, \tag{9.6}$$

$$\mathsf{P}(\rho(n^{-1}S_n, \Phi(s)) > \delta) \leq \exp(-n(s + \gamma)), \qquad n \to \infty, \tag{9.7}$$

where $\Phi(s) = (z\colon z \in E^l\colon L(z) \leq s)$ is a compact.

Proof of Theorem 9.1. Let us denote

$$G_n^+(\alpha) = n^{-1} \ln \sup_x \mathsf{E}_x \exp(\alpha S_n), \qquad G_n^-(\alpha) = n^{-1} \ln \inf_x \mathsf{E}_x \exp(\alpha S_n).$$

For $n, k > 0$ we obtain

$$\begin{aligned} G_{n+k}^+(\alpha) &\leq \frac{n}{n+k} G_n^+(\alpha) + \frac{k}{n+k} G_k^+(\alpha), \\ G_{n+k}^-(\alpha) &\geq \frac{n}{n+k} G_n^-(\alpha) + \frac{k}{n+k} G_k^-(\alpha). \end{aligned}$$

By virtue of subadditive functions properties (see, for example, Stroock (1984), Lemma 3.11) there exist the limits

$$G^+(\alpha) = \lim G_n^+(\alpha), \qquad G^-(\alpha) = \lim G_n^-(\alpha).$$

Indeed, let $m > 0$, $n = q_n m + r_n$ for any n $(0 \leq r_n < m)$. Then

$$G_m^+ = G_{q_n + r_n}^+ \leq n^{-1} q_n m G_{q_n m}^+ + n^{-1} r_n G_{r_n}^+ \leq n^{-1} q_n m G_m^+ + n^{-1} r_n G_{r_n}^+.$$

It follows that

$$\limsup_{n \to \infty} G_n^+ \leq G_m^+.$$

Thus,

$$\limsup_{n \to \infty} G_n^+ \leq \limsup_{m \to \infty} G_m^+.$$

By virtue of conditions (9.4) and (9.5)

$$G_n^+(\alpha) \leq n^{-1} \ln C_n + G_n^-(\alpha),$$

so that $G^+(\alpha) \le G^-(\alpha)$, but, on the other hand, $G^-(\alpha) \le G^+(\alpha)$. So $G^+(\alpha) = G^-(\alpha)$ and the uniform limit (9.1) does exist with $G(\alpha) = G^{\pm}(\alpha)$. Theorem 9.1 is proved.

Proof of Theorem 9.2. Inequality (9.7) follows from Theorem 9.1 and Theorem 5.1.1 from Freidlin and Wentzell (1984). Theorem 5.1.1 does not require the differentiability of G. The compactness of $\Phi(s)$ follows from the properties of function $L(z)$: it is lower semicontinuous and is equal to $+\infty$ outside some compact.

Let us prove inequality (9.6). It is convenient to divide the proof into steps.

A. Denote $x = (X_{-\infty}^0) \in Z_-$, $\mathbf{0} = (0,\ldots,0,\ldots)$. By virtue of conditions (9.4)–(9.5) for any $x, y \in Z_-$

$$C_n^{-1} \le \frac{\mathsf{P}_x(|n^{-1}S_n - z| < \delta)}{\mathsf{P}_y(|n^{-1}S_n - z| < \delta)} \le C_n.$$

Hence,

$$n^{-1}|\ln \mathsf{P}_x(|n^{-1}S_n - z| < \delta) - P_0(|n^{-1}S_n - z| < \delta)| \le n^{-1}C_n \to 0, \quad n \to \infty.$$

So it is sufficient to prove the estimate for $x = \mathbf{0}$ only.

B. Let $\delta > 0$, $k > 0$. Let us build the sequence of random vectors $(\hat{X}_n, n \in Z)$ on some extension of our probability space in such a way that $\hat{X}_0 = \hat{X}_{-1} = \ldots = 0$, $\hat{X}_1 = \hat{X}_1, \ldots, \hat{X}_k = X_k$, and vectors $(\hat{X}_{ik+1}, \ldots, \hat{X}_{(i+1)k})$, $i \ge 0$, are independent and identically distributed. Denote by $\hat{\mathsf{P}}$ the probability measure on this new probability space, by $\hat{\mathsf{E}}$—expectation with respect to $\hat{\mathsf{P}}$, and $\hat{S}_n = \sum_{k=1}^n \varphi(\hat{X}_k)$.

By virtue of condition (9.4) for any $\Gamma_{mk} \in \mathcal{B}^{mk}$ we obtain

$$C_k^{-m} \le \frac{\hat{\mathsf{P}}(\hat{X}_1^{mk} \in \Gamma_{mk})}{\mathsf{P}(X_1^{mk} \in \Gamma_{mk})} \le C_k^m. \tag{9.8}$$

Indeed,

$$C_k^{-1} \le \frac{\mathsf{P}(X_{(m-1)k+1}^{mk} \in dy | X_{-\infty}^{(m-1)k})}{\hat{\mathsf{P}}(\hat{X}_{(m-1)k+1}^{mk} \in dy)} \le C_k, \tag{9.9}$$

so that (9.8) follows from (9.9) by induction. Note that any point mk, $m \in Z_+$, is the regeneration point for the process $\hat{X}$ under the measure $\hat{\mathsf{P}}$.

(9.8) implies the two-sided estimate

$$C_k^{-m} \le \frac{\mathsf{P}(\rho(n^{-1}S_n, z) < \delta)}{\hat{\mathsf{P}}(\rho(n^{-1}S_n, z) < \delta)} \le C_k^m.$$

Hence,

$$\mathsf{P}(\rho(n^{-1}S_n, z) < \delta) \ge C_m^{-k}\hat{\mathsf{P}}(\rho(n^{-1}S_n, z) < \delta).$$

By virtue of the same reasons we conclude

$$\mathsf{P}(\rho(n^{-1}S_n, z) < \delta) \ge C_m^{-[n/k]-1}\mathsf{P}(\rho(n^{-1}S_n, z) < \delta). \tag{9.10}$$

So

$$\liminf_{n\to\infty} n^{-1}\ln \mathsf{P}(\rho(n^{-1}S_n, z) < \delta) \geq \liminf_{n\to\infty} n^{-1}\ln \hat{\mathsf{P}}(\rho(n^{-1}\hat{S}_n, z) < \delta) - k^{-1}\ln C_k.$$

By virtue of condition (9.5) the value k may be chosen in such a way that $k^{-1}\ln C_k < \gamma/2$.

C. Let us denote

$$\hat{G}^k(\alpha) = \lim_{n\to\infty} n^{-1}\ln \hat{\mathsf{E}}\exp(\alpha\hat{S}_n), \tag{9.11}$$

Let us show that the limit in (9.11) exists and function $\hat{G}^k(\alpha)$ is differentiable. Indeed,

$$\begin{aligned}
\limsup_{n\to\infty} n^{-1}\ln \hat{\mathsf{E}}\exp(\alpha\hat{S}_n) &= \limsup_{n\to\infty}(q_nk + r_n)^{-1}\ln \hat{\mathsf{E}}\exp(\alpha\hat{S}_{q_nk} + \alpha\hat{S}_n^{q_nk+1}) \\
&\leq \limsup_{n\to\infty}(q_nk + r_n)^{-1}\ln \hat{\mathsf{E}}\exp(\alpha\hat{S}_{q_nk} + |\alpha|\,||\varphi||k) \\
&= \limsup_{q_n\to\infty}(q_nk)^{-1}\ln \hat{\mathsf{E}}\exp(\alpha\hat{S}_{q_nk}) \\
&\quad + \limsup_{q_n\to\infty}(q_n + r_n)^{-1}|\alpha|\,||\varphi||k \\
&= k^{-1}\ln \hat{\mathsf{E}}\exp(\alpha\hat{S}_k),
\end{aligned}$$

because $\hat{S}_{qk}$ is the sum of q independent identically distributed random values with the same distribution as $\hat{S}_k$. Similarly the lower estimate is obtained:

$$\begin{aligned}
\liminf_{n\to\infty} n^{-1}\ln \hat{\mathsf{E}}\exp(\alpha\hat{S}_n) &\geq \liminf_{n\to\infty}(q_nk + r_n)^{-1}\ln \hat{\mathsf{E}}\exp(\alpha\hat{S}_{q_nk} - |\alpha|\,||\varphi||k) \\
&= k^{-1}\ln \hat{\mathsf{E}}\exp(\alpha\hat{S}_k).
\end{aligned}$$

D. Let us denote $\hat{L}^k(z) = \sup(\alpha z - \hat{G}^k(\alpha))$. Since $\hat{G}^k$ is differentiable then the large deviations principle holds for the sums $\hat{S}_n$ (see Freidlin and Wentzell (1984), Theorems 5.1.1 and 5.1.2). Hence,

$$\hat{\mathsf{P}}(\rho(n^{-1}\hat{S}_n, z) < \delta) \geq \exp(-n(\hat{L}^k(z) + \gamma)), \qquad n\to\infty. \tag{9.12}$$

We may assume $L(z) < \infty$ (else inequality (9.6) is trivial). Moreover, we may assume that z belongs to the interior A_0 of the set $A = (y\colon\ L(y) < \infty)$ w.r.t. its linear hull. Indeed, for any $z \in A$ the equality holds

$$L(z) = \liminf_{y\to z,\ y\in A_0} L(y)$$

(see Rockafellar (1970)).

For any $z \in A_0$ there exists such α that $L(z) = \alpha z - G(\alpha)$ (see Rockafellar (1970)).

E. Suppose that function G is differentiable in the given point α. Then $\nabla G(\alpha) = z$. Hence, by virtue of subgradient convergence theorem (see Rockafellar (1970)) $\nabla\hat{G}^k(\alpha) \to \nabla G(\alpha) = z$, since $\hat{G}^k \to G$, $k\to\infty$. This implies the convergence $\hat{L}^k(z) \to L(z)$. Indeed, denote $\nabla\hat{G}^k(\alpha) = z_k$. Then $\hat{L}^k(z_k) = \alpha z_k - \hat{G}^k(\alpha)$, so that $\lim_{k\to\infty}\hat{L}^k(z_k) = \lim_{k\to\infty}\alpha z_k - \lim_{k\to\infty}\hat{G}^k(\alpha) = \alpha z - G(\alpha) = L(z)$. Hence, inequalities (9.11) with z_k and (9.10) as $k\to\infty$ imply (9.6).

F. Suppose now that G is not differentiable in the point α. Hence, G is not differentiable along some body cone K of directions. The cone K is supposed to be maximal one. Then G is differentiable along $K^{\perp}$. Let us denote by K' the cone of subgradients to the graph of G in the point α which correspond to directions from K. We may assume that correspondent coordinates of z belong to the interior of K'. Let us denote these coordinates by z' and complement ones by z''. Let $z = (z', z'')$ and $\alpha = (\alpha', \alpha'')$ in the adjoint space.

Let us build sequences $\alpha_k = (\alpha'_k, \alpha''_k)$ and $z_k = (z', z'')$ by the following way.

(1) $z'_k = z'$, $\alpha''_k = \alpha''$.

(2) $\alpha'_k = \arg\max_{\alpha'} (\alpha' z'_k - \hat{G}^k(\alpha', \alpha''_k))$.

(This max is attained for $k \geq k_\alpha$).

(3) $z''_k = \nabla_{\alpha''} \hat{G}^k(\alpha'_k, \alpha'') = \arg\max_{z''} (\alpha'' z'' - \hat{L}^k(z'_k, z''))$.

G. Let us show that $\alpha_k \to \alpha$ and $z_k \to z$. We have $\alpha''_k = \alpha''$ and $z''_k = z'$. Now, since $\hat{G}^k \to G$ then $\alpha'_k \to \alpha'$. Finally, function G is differentiable in α'' in the point α. Thus, by virtue of subgradients convergence theorem (Rockafellar (1970)) and the equality $\lim \alpha'_k = \alpha'$ we obtain

$$z''_k \to \nabla_{\alpha''} G(\alpha', \alpha'') = z''.$$

Hence, $\alpha_k \to \alpha$, $z_k \to z$. So

$$\lim_{k\to\infty} \hat{L}^k(z_k) = \lim_{k\to\infty} \alpha_k z_k - \lim_{k\to\infty} \hat{G}^k(\alpha_k) = \alpha z - G(\alpha) = L(z).$$

This implies (9.6) in the same way as in part D. Theorem 9.2 is proved.

Example 9.1. Let X_n take their values in the torus $[0, 2\pi)$ and satisfy the equation

$$X_{n+1} = \left\{ \sum_{i=0}^{\infty} a_i X_{n-i} + W_{n+1} \right\},$$

where W_n are i.i.d. random variables in $[0,1)$ with a positive smooth density $p(y)$, $\{a\} = a - [a]$ where $[a]$ is the integer part of the value a. Assume $\sum_i |a_i| < \infty$. We have

$$\mathsf{P}(X_{n+1} \in \mathrm{d}y_1 | X_{-\infty}^n)/\,\mathrm{d}y_1 = p\left(\left\{ \sum_{i=0}^{\infty} a_i X_{n-i} + y_1 \right\} \right),$$

$$\mathsf{P}(X_{n+2} \in \mathrm{d}y_2 | X_{-\infty}^n, X_{n+1} = y_1)/\,\mathrm{d}y_2 = p\left(\left\{ \sum_{i=0}^{\infty} a_{i+1} X_{n-i} + y_1 \right\} \right),$$

$$\begin{aligned} &\mathsf{P}(X_{n+3} \in \mathrm{d}y_3 | X_{-\infty}^n, X_{n+1} = y_1, X_{n+2} = y_2)/\,\mathrm{d}y_3 \\ &\quad = p\left(\left\{ \sum_{i=0}^{\infty} a_{i+2} X_{n-i} + a_1 y_1 + a_0 y_2 + y_3 \right\} \right), \end{aligned}$$

$$\mathsf{P}(X_{n+3} \in \mathrm{d}y_3 | X^n_{-\infty}, X_{n+1} = y_1, X_{n+2} = y_2)/\,\mathrm{d}y_3$$
$$= p\left(\left\{\sum_{i=0}^{\infty} a_{i+2}X_{n-i} + a_1y_1 + a_0y_2 + y_3\right\}\right), \ldots$$

$$\mathsf{P}(X_{n+k} \in \mathrm{d}y_k | X^n_{-\infty}, X_{n+1} = y_1, \ldots, X_{n+k-1} = y_{k-1})/\,\mathrm{d}y_k$$
$$= p\left(\left\{\sum_{i=0}^{\infty} a_{i+2}X_{n-i} + a_{i+k-2}y_1 + \ldots + y_k\right\}\right), \ldots$$

Denote

$$p(X, y_1, \ldots, y_k; y_{k+1}) = \mathsf{P}(X_{k+1} \in \mathrm{d}y_{k+1} | X^0_{-\infty} = X, X_1 = y_1, \ldots, X_k = y_k)/\,\mathrm{d}y_k,$$

$X \in Z_-$. Then for any $X, Y \in Z_-$

$$p(X, y_1) \ldots p(X, y_1, \ldots, y_k; y_{k+1}) \leq \prod_{i=1}^{k} C_i p(Y, y_1, \ldots, y_i; y_{i+1}),$$

where

$$C_i = \sup_{X,Y,y_1,\ldots,y_i} \frac{p(X, y_1, \ldots, y_i; y)}{p(Y, y_1, \ldots, y_i, y)} \leq \sup_{\substack{x,\, y \in [0,1] \\ |x-y| \leq \sum_{k>i-1} |a_k|}} \frac{p(x)}{p(y)}, \qquad i \geq 1.$$

Suppose p satisfies Lipschitz condition

$$|p(x) - p(y)| \leq L|x - y|.$$

Then,

$$0 \leq C_i - 1 \leq C \sum_{k=i}^{\infty} |a_k|.$$

Condition (9.3) is satisfied if the value $\prod_{\infty}^{i=1} C_i$ is finite, that is, if $\sum_{j=1}^{\infty} j|a_j| < \infty$. Conditions (9.4)–(9.5) are satisfied if

$$\lim_{n\to\infty} n^{-1} \ln \prod_{i=1}^{n} C_i = 0,$$

that is, if $\sum_{j=1}^{\infty} |a_j| < \infty$. Note that the latter condition was assumed to hold at the beginning of the example.

Example 9.2. Let X_n satisfy the equation

$$X_{n+1} = \{F(X^n_{-\infty}) + W_n\},$$

W_n are i.i.d. random variables in $[0, 1)$ with a smooth positive density p which satisfies the Lipschitz condition. Let us denote

$$c_k = \sup_{x_i \in R, X, Y \in Z_-} |F(x_0, x_{-1}, \ldots, x_{-k}, Y) - F(x_0, x_{-1}, \ldots, x_{-k}, X)|.$$

Then conditions (9.4)–(9.5) are satisfied if $\sum_{j=1}^{\infty} |c_j| < \infty$. It follows from the same considerations as in the previous example.

9.2. NON-COMPACT NON-SINGULAR CASE

Here the stationary ergodic process $(X_n, -\infty < n < +\infty)$ with values in E^d is considered, and

$$S_n = \sum_{i=1}^{n} \varphi(X_i),$$

where φ is a Borel function (not necessary bounded), φ: $E^d \to E^l$. Large deviations are studied under stability and local non-singularity conditions similar to ones in the Markovian case.

We assume that the conditional distribution of X_n with respect to $F_{n-1} = \sigma(X_{n-1}, X_{n-2}, \ldots)$ does not depend on n.

Under this assumption the infinite-dimensional process $X^n_{-\infty} = (X_n, X_{n-1}, \ldots)$ is the homogeneous Markov one. Nevertheless, it is difficult to use the Frobenius-type theorem on spectral radius because of the strong degeneracy of the correspondent operator.

Let us state stability and non-degeneracy conditions. Denote $Z_- = (0, -1, -2, \ldots)$, $Z_+ = (1, 2, \ldots)$. We suppose there exists the sequence $(K^m, m \geq 0)$ of measurable sets from E^{Z_-} such that $K^m \uparrow (E^d)^{Z_-}$ and

(A) for any $m \geq 0$ there exist finite functions $h^+_m(x) \geq h^-_m(x) > 0$, $x = (x_0, x_{-1}, \ldots) \in (E^d)^{Z_-}$ such that for any $a > 0$, $m \geq m_a$, $x \in K^m$

$$h^-_m(x) \leq \mathsf{E}(\exp(a\tau_m) | X^0_{-\infty} = x) \leq h^+_m(x),$$

where $\tau_m = \inf(k \geq 1\colon X^k_{-\infty} \in K^m)$ we can always choose $h^-_m \equiv 1$;

(A_φ) for any $m \geq 0$ there exist such finite functions $h^+_m(x) \geq h^-_m(x) > 0$, $x = (x_0, x_{-1}, \ldots) \in (E^d)^{Z_-}$ that for any $a > 0$, $m \geq m_a$, $x \in K^m$

$$h^-_m(x) \leq \mathsf{E}\left(\exp\left(\alpha \sum_{k=1}^{\tau_m} \varphi(X_k)\right) | X^0_{-\infty} = x\right) \leq h^+_m(x);$$

(B) for any $a > 0$, $m \geq m_a$, $n \geq 0$, $x \in (E^d)^{Z_-}$

$$C^{-1}_m \leq \mathsf{E}(h^-_m(X^n_{-\infty}) | X^0_{-\infty} = x) \leq \mathsf{E}(h^+_m(X^n_{-\infty}) | X^0_{-\infty} = x) \leq C_m$$

$(C_m > 0)$, and $\lim_{m \to \infty} C_m = 1$, where $h^{\mp}$ are functions from (A) or (A_φ);

(C)

$$\lim_{m \to \infty} \inf_{x \in K^m} \mathsf{P}(X^1_{-\infty} \in K^m | X^0_{-\infty} = x) = 1;$$

(D_b) for any $m \geq 0$ there exists such a constant $L_m > 0$ that for any $x, y \in K^m$ and $\Gamma \in (E^d)^{Z_+}$

$$\mathsf{P}(X^\infty_1 \in \Gamma | X^0_{-\infty} = x) \leq L_m \mathsf{P}(X^\infty_1 \in \Gamma | X^0_{-\infty} = y);$$

(D) for any $m \geq 0$ there exist such constants $L_m(N)$, $N \geq 1$, that

$$\lim_{N\to\infty} N^{-1} \ln L_m(N) = 0$$

and for any $x, y \in K^m$ and $\Gamma_N \in \mathcal{B}((E^d)^N)$

$$\mathsf{P}(X_1^N \in \Gamma | X_{-\infty}^0 = x) \leq L_m(N) \mathsf{P}(X_1^N \in \Gamma | X_{-\infty}^0 = y).$$

In the beginning of the chapter we studied the special case of this situation when one may take $K^0 = (E^d)^{Z_-}$ and φ is bounded. Below we consider some simple examples of processes which satisfy conditions (A)–(D) but do not satisfy conditions of the first part of the chapter. In the Markovian case conditions (A)–(C) looks similar with $K^m \subset E^d$. (D) and (D_b) in the Markovian case are rather strong Doeblin-type conditions. Apparently it shows that these assumptions may be weakened.

Remark 9.1. In Orey and Pelikan (1988) condition $(R\text{–}M)$ is introduced which is similar to our condition (D_b). (D_b) may be replaced by $(R\text{–}M)$ here as well.

Theorem 9.3. Let conditions (A), (A_φ), (B), (C) and (D) be satisfied. Then for any $x \in (E^d)^{Z_-}$ and $\alpha \in E^l$ there exists the finite limit

$$G(\alpha) = \lim_{n\to\infty} n^{-1} \ln \mathsf{E}_x \exp(\alpha S_n), \tag{9.13}$$

where function G does not depend on x and is convex in α.

Let us denote $L(\beta) = \sup(\alpha\beta - G(\alpha))$.

Theorem 9.4. Let conditions of Theorem 9.3 be satisfied. Then $nL(\beta)$ is the action functional for the family (S_n/n) as $n \to \infty$, that is, for any $z \in E^l$, $s > 0$

$$\liminf_{\delta\to 0} \liminf_{n\to\infty} n^{-1} \ln \mathsf{P}_x(\rho(n^{-1}S_n, z) < \delta) \geq -L(z), \tag{9.14}$$

$$\limsup_{\delta\to 0} \limsup_{n\to\infty} n^{-1} \ln \mathsf{P}_x(\rho(n^{-1}S_n, \Phi(s)) > \delta) \leq -s, \tag{9.15}$$

where $\Phi(s) = (z \in E^l\colon L(z) \leq s)$ is a compact in E^l. The limits in (9.14) and (9.15) are uniform in $x \in K^m$ for any m.

Remark 9.2. Condition (A_φ) is used for φ unbounded. If φ is bounded then (A_φ) follows from (A).

Preliminary results

Let us denote $\sigma_m = \sigma_m^1 = \inf(n \geq 1\colon X_{n-1} \in K^m, X_n \in K^m)$, $\sigma_m^{i+1} = \vartheta_{\sigma_m}\sigma_m^i$ (ϑ is a shift operator).

Lemma 9.1. Let conditions (A_φ), (B) and (C) be satisfied. Then for any $a > 0$, $|\alpha| < a$, $x \in (E^d)^{Z_-}$ the following inequalities hold:

$$C^{-1}h_{m,2a}^{-}(x)^{1/2} \leq \mathsf{E}\left(\exp\left(\alpha \sum_{k=1}^{\sigma_m} \varphi(X_k)\right) | X_{-\infty}^0 = x\right) \leq C h_{m,2a}^{+}(x)^{1/2}. \tag{9.16}$$

Proof follows from assumptions similar to the Markovian case (see previous chapter).

Let us denote $l(n) = \sup(k = 0, 1, \dots : \sigma_m^k \le n)$.

Lemma 9.2. (1) Let conditions (A), (B) and (C) be satisfied. Then for any $\delta > 0$ and $C > 0$

$$\mathsf{P}(|n^{-1}l(n) - 1| > \delta) \le \exp(-Cn), \tag{9.17}$$

if m is large enough and $n \ge n(\delta, x)$.

(2) Let conditions (A), (A_φ), (C) and (D) be satisfied. Then for any $\delta > 0$ and $C > 0$

$$\mathsf{P}(|S_{\sigma_m^n} - S_n| > \delta n | X_{-\infty}^0 = x) \le \exp(-Cn), \tag{9.18}$$

if m is large enough and $n \ge n(\delta, x)$.

Proof follows from Lemma 9.1 similar to the Markovian case.

Let us denote $\hat{S}_n^m = \sum_{k=1}^{\sigma_m^n} \varphi(X_k) \mathrm{I}(X_{-\infty}^k \in K^m)$.

Lemma 9.3. Let conditions (A), (A_φ), (C) and (D) be satisfied. Then for any $\delta > 0$ and $C > 0$

$$\mathsf{P}(|S_{\sigma_m^n} - \hat{S}_n^m| > \delta n | X_{-\infty}^0 = x) \le \exp(-Cn), \tag{9.19}$$

if m is large enough and $n \ge n(\delta, x)$.

Proof. By virtue of Chebyshov's inequality

$$\begin{aligned} &\mathsf{P}(\pm(S_{\sigma_m^n} - \hat{S}_n^m) > \delta n | X_{-\infty}^0 = x) \\ &\quad\le \exp(-\lambda\delta n)\, \mathsf{E}_x \exp\left(\pm\lambda \sum_{k=1}^{\sigma_m^n} \varphi(X_k) \mathrm{I}(X_{-\infty}^k \in K^m)\right), \end{aligned} \tag{9.20}$$

where E_x means the same as $\mathsf{E}(\dots | X_{-\infty}^0 = x)$. Denote

$$\xi_m^n = \pm \sum_{k=1}^{\sigma_m^n} \varphi(X_k) \mathrm{I}(X_{-\infty}^k \in K^m).$$

Then by virtue of Lemma 9.1 $\mathsf{E}(\exp(\lambda \xi_m^n) | F_{\sigma_m^{n-1}}) \le L_m$ for any n and any λ if $m \ge m(\lambda, x)$, and, moreover, $\mathsf{P}(\xi_m^n \ne 0 | F_{\sigma_m^{n-1}}) \le p_m \ll 1$ as $m \to \infty$. Thus, by induction we obtain

$$\begin{aligned} &\mathsf{E}_x \exp\left(\pm\lambda \sum_{k=1}^{\sigma_m^n} \varphi(X_k) \mathrm{I}(X_{-\infty}^k \in K^m)\right) \\ &\quad= \mathsf{E}_x \exp\left(\pm\lambda \sum_{k=1}^{\sigma_m^n} \xi_k^m\right) = \mathsf{E}_x \exp\left(\pm\lambda \sum_{k=1}^{\sigma_m^{n-1}} \xi_k^m\right) \mathsf{E}(\exp(\pm\lambda \xi_n^m) | F_{\sigma_m^{n-1}}) \\ &\quad\le \mathsf{E}_x \exp\left(\pm\lambda \sum_{k=1}^{\sigma_m^{n-1}} \xi_k^m\right) \{1 + \mathsf{E}(\exp(\pm\lambda \xi_n^m) | F_{\sigma_m^{n-1}}) \mathsf{P}(\xi_n^m \ne 0)\} \\ &\quad\le \mathsf{E}_x \exp\left(\pm\lambda \sum_{k=1}^{\sigma_m^{n-1}} \xi_k^m\right) \{1 + L_m p_m\} \le \dots \le (1 + L_m p_m)^n. \end{aligned} \tag{9.21}$$

Since $\lim_{m\to\infty} L_m p_m = 0$ for any λ then we may put $\lambda = 2C/\delta$ and for m large enough (9.20) and (9.21) imply (9.19). Lemma 9.3 is proved.

Proof of Theorem 9.3. In 'compact' case when one may put $K^m = (E^d)^Z$- in all conditions (A)–(D) the existence of the limit follows from the subadditive functions. In our more general case the idea is to reduce the proof to compact case again. Let us divide the proof into several steps.

I. By virtue of Lemmas 9.2 and 9.3 for any $\delta > 0$ and $C > 0$ there exist such m and n_0 that

$$\mathsf{P}\left(\sup_{n\geq n_0} n^{-1}|S_n - \hat{S}_n^m| > \delta/2\right) \leq \exp(-Cn). \tag{9.22}$$

II. Inequality (9.22) implies that for any $\delta > 0$, $C > 0$ and $a > 0$ there exists such m_0 that for $m \geq m_0$, $|\alpha| \leq a$, $p > 1$, $q = p/(p-1)$ and $n \geq n(x)$

$$(\mathsf{E}_x \exp(p^{-1}\alpha\hat{S}_n^m))^p(1+\delta)^{-n} \leq \mathsf{E}_x \exp(\alpha S_n) \leq (\mathsf{E}_x \exp(p\alpha\hat{S}_n^m))^{1/p}(1+\delta)^n. \tag{9.23}$$

Indeed, by virtue of the Hölder inequality

$$\begin{aligned}\mathsf{E}_x \exp(\alpha S_n) &\leq \mathsf{E}_x \exp(\alpha\hat{S}_n^m + \alpha(S_n - \hat{S}_n^m)) \\ &\leq (\mathsf{E}_x \exp(p\alpha\hat{S}_n^m))^{1/p}(\mathsf{E}_x \exp(q\alpha(S_n - \hat{S}_n^m)))^{1/q}.\end{aligned}$$

For $m \geq m(qa, \delta)$ we obtain by virtue of (9.22)

$$\mathsf{E}_x \exp(q\alpha(S_n - \hat{S}_n^m)) \leq (1 + p_m)^n,$$

where $p_m > 0$ is arbitrary small as $m \to \infty$. Then

$$\mathsf{E}_x \exp(\alpha S_n) \leq (\mathsf{E}_x \exp(p\alpha\hat{S}_n^m))^{1/p}(1+\delta)^n.$$

Similarly we obtain the lower estimate. Namely, by virtue of the Hölder inequality

$$\mathsf{E}_x \exp(\alpha S_n) \geq (\mathsf{E}_x \exp(p^{-1}\alpha\hat{S}_n^m))^p(\mathsf{E}_x \exp(qp^{-1}\alpha(S_n - \hat{S}_n^m)))^{-p/q},$$

etc.

III. Consider the process $Y_n = X_{\sigma_m^n}$, $n \geq 0$, $Y_n = X_n$, $n < 0$, and the family of σ-fields $\hat{F}_n = F_{\sigma_m^n}$. For any $n > 0$, $Y_n \in K^m$. Let us show that for any $n > 0$, $k \geq 1$, $x, y \in K^m$ the following assertions hold:

$$\mathsf{P}(Y_{n+1}^{n+k} \in \Gamma_k | X_{-\infty}^{\sigma_m^n} = x) \leq L_m'(k)\,\mathsf{P}(Y_{n+1}^{n+k} \in \Gamma_k | X_{-\infty}^{\sigma_m^n} = y), \tag{9.24}$$

$$\lim_{k\to\infty} k^{-1}\ln L_m'(k) = 0, \tag{9.25}$$

that the process $(Y_n)_{n\geq 1}$ satisfies the 'compact' or 'uniform' analog of condition (D), since it is not necessary to consider the case $x, y \notin K^m$ (because $X_{\sigma_m^n}^{-\infty} \in K^m$, $n \geq 1$).

Indeed, by virtue of conditions (D) and (C)

$$\begin{aligned} \mathsf{P}(Y_{n+1}^{n+k} \in \Gamma_k | X_{-\infty}^{\sigma_m^n} = x) &= \sum_{i=k}^{\infty} \mathsf{E}(\mathrm{I}(\sigma_m^{n+k} - \sigma_m^n = i)\,\mathrm{I}(Y_{n+1}^{n+k} \in \Gamma_k)|X_{-\infty}^{\sigma_m^n} = x) \\ &\le \sum_{i=k}^{\infty} L_m(i) p_m^i \mathsf{E}(\mathrm{I}(\sigma_m^{n+k} - \sigma_m^n = i)\,\mathrm{I}(Y_{n+1}^{n+k} \in \Gamma_k)|X_{-\infty}^{\sigma_m^n} = y) \\ &\equiv L'_m(k)\,\mathsf{P}(Y_{n+1}^{n+k} \in \Gamma_k | X_{-\infty}^{\sigma_m^n} = y), \end{aligned} \tag{9.26}$$

where $p_m = 1 - \sup_{x \in K^m} \mathsf{P}_x(X_{-\infty}^1 \in K^m)$ is arbitrary small. By virtue of condition (D) $\lim_{N\to\infty} N^{-1} \ln L_m(N) = 0$. In other words, function $L_m(N)$ grows in N slower than any exponent. Hence, the sum in (9.26) converges and $L'_m(k) < \infty$.

Let us check condition (9.25). We have

$$\begin{aligned} \limsup_{k\to\infty} k^{-1} \ln L'_m(k) &= \limsup_{k\to\infty} k^{-1} \ln \sum_{i\ge k} L_m(i) p_m^{i-k} \\ &\le \limsup_{k\to\infty} k^{-1} \left(\ln L_m(k) + \ln\Big(1 + \sum_{i>k} L_m(i) L_m(k)^{-1} p_m^{i-k}\Big)\right) \\ &\le \lim_{k\to\infty} k^{-1} \ln L_m(k) + \limsup_{k\to\infty} k^{-1} \sum_{i>k} L_m(i) L_m(k)^{-1} p_m^{i-k}. \end{aligned}$$

By virtue of condition (D) the sum $\sum_{i>k} L_m(i) L_m(k)^{-1} p_m^{i-k} < \infty$, and thus,

$$\limsup_{k\to\infty} k^{-1} \ln L'_m(k) = 0.$$

This implies assertions (9.24)–(9.25).

IV. By virtue of (9.24)–(9.25) the limit

$$\hat{G}^m(\alpha) = \lim_{n\to\infty} n^{-1} \ln \mathsf{E}_x \exp(\alpha \hat{S}_n^m), \tag{9.27}$$

for the sums $\hat{S}_n^m = \sum_{k=1}^n Y_k$ exists for any $a > 0$, $|\alpha| \le a$, $m \ge m(a)$ and this limit does not depend on x and is convex in α (cf. with the first part of the book). Indeed, denote $G_n^{m+}(\alpha) = n^{-1} \sup_x \ln \mathsf{E}_x \exp(\alpha \hat{S}_n^m)$, $G_n^{m-}(\alpha) = n^{-1} \inf_x \ln \mathsf{E}_x \exp(\alpha \hat{S}_n^m)$, $G_n(\alpha) = n^{-1} \ln \mathsf{E}_x \exp(\alpha S_n)$. Since $Y_i \in K^m$, $i \ge 1$, then by virtue of (9.24)

$$G_n^{m+}(\alpha) \le G_n^{m-}(\alpha) + n^{-1} \ln L_m([n/m] + 1).$$

This inequality together with (9.25) implies that for $m \ge m(|\alpha|)$

$$\limsup_{n\to\infty} G_n^{m+}(\alpha) = \liminf_{n\to\infty} G_n^{m-}(\alpha),$$

and hence the limit (9.27) does exist.

V. Finally, (9.27) and (9.22) imply

$$\limsup_{n\to\infty} G_n(\alpha) \le \liminf_{m\to\infty} \hat{G}^m(\alpha),$$

$$\liminf_{n\to\infty} G_n(\alpha) \ge \limsup_{m\to\infty} \hat{G}^m(\alpha).$$

This proves the existence of the limit (9.13), and moreover,

$$\lim_{n\to\infty} G_n(\alpha) = \lim_{m\to\infty} \hat{G}^m(\alpha) = \lim_{m\to\infty}\lim_{n\to\infty} \hat{G}^m_n(\alpha) \tag{9.28}$$

(this assertion is used below in the proof of Theorem 9.4). Theorem 9.3 is proved.

Proof of Theorem 9.4. The upper bound (9.15) follows from Theorem 9.3 and Theorem 5.1.1 from Freidlin and Wentzell (1984) in which the differentiability of the limit G is not assumed. the compactness of the set $\Phi(s)$ follows from the properties of function L.

Proof of the lower bound (9.14) follows the same scheme as in the 'compact' case but with some new tricks only. So we shall not repeat all steps.

Let us denote $\hat{L}(z) = \sup(\alpha z - \hat{G}(\alpha))$. By virtue of inequality (9.22) and lower estimate (9.6) the following inequality holds true:

$$\liminf_{n\to\infty} n^{-1} \ln \mathsf{P}(|n^{-1}S_n - z| < \delta) \geq -\hat{L}^m(z'), \tag{9.29}$$

if $|z - z'| < \delta/2$ and m is large enough. Let $A = (z\colon\ L(z) < \infty)$, $A_0 = A^{\text{int}}$ is the interior of A with respect to the linear hull. Let us take $z' \in A_0$. Then the same considerations as in the proof of Theorem 9.2 with the help of equalities (9.13) and (9.28) show that for $z' \in A_0$ one may choose such $z^m_n \to z'$, $m, n \to \infty$, that

$$\lim_{m,n\to\infty} \hat{L}^m(z^m_n) = L(z'). \tag{9.30}$$

Assertions (9.29) and (9.30) together with the lower semicontinuity of $L(z)$ imply the desired estimate (9.14). Theorem 9.4 is proved.

Remark 9.3. We may say that in some sense the logarithmic asymptotics of the type (9.14) and (9.15) may be established with our method for such stationary processes which may be well approximated by Markov ones with good enough properties of stability and non-degeneracy.

Example 9.3. Consider the one-dimensional process

$$X_{n+1} = X_n + a(X^{-\infty}_n) + W_{n+1},$$

where W_n are i.i.d. random variables, $\mathsf{E}W_n = 0$, $\mathsf{E}\exp(\alpha|W_n|) < \infty$ for any α and W_n has a positive smooth distribution density. Let function a satisfies conditions

$$|a(x)| \leq L(1 + |x_0|) \tag{9.31}$$

($x = (x_0, x_{-1}, \ldots)$) and

$$\lim_{x_0\to\infty} (|x_0 + a(x)| - |x_0|) = -\infty \tag{9.32}$$

(uniformly in $(x_{-1}, x_{-2}, \ldots)$). Then easy calculations show that conditions (A), (B), and (C) are satisfied with $K^m = (x \in (E^d)^{Z_-}\colon |x_0| \leq m)$, $h^{\mp}_m(x) = \exp(\pm\varepsilon(|x_0| - m))$

and some $\varepsilon = \varepsilon(a) > 0$. In other words, the process must satisfy stability and non-degeneracy conditions in its last coordinate uniformly in other ones in the same way as in the Markovian case.

Example 9.4. Consider a one-dimensional stationary linear-regular (see Ibragimov and Linnik (1971)) autoregression process

$$X_{n+1} = \sum_{i\geq 0} a_i X_{n-i} + W_{n+1}$$

under the same assumptions on W_n as in the previous example. Let $\sup_i |a_i|\lambda^{-i}\hat{a} < 1$, $\lambda \in (0,1)$, $0 < \hat{a}/(1-\lambda) < 1$. These assumptions are very rough but now we only want to construct an example not similar to 9.3. By virtue of the linear regularity one may use the implicit formula for X_n. This formula particularly implies that $\sum_{i<0} |X_i| q^i < \infty$ almost everywhere for any $q \in (0,1)$. Let us fix $q < 1 - \hat{a}(1-\lambda)^{-1}$ and let $\| x \| = \sum_{i\geq 0} |x_{-i}| q^i$, $K^m = (x = (x_0, x_{-1}, \ldots) \in R^{Z_-} \colon \sum_{i\geq 0} |x_{-i}| q^i \leq m)$. Condition (C) is checked trivially. Let $h^{\pm}(x) = \exp(\varepsilon \| x \|)$, $\varepsilon = \varepsilon(a) > 0$. Let us show that condition (A) is also fulfilled for appropriate constants. For $n < \tau_m$ we have

$$\begin{aligned} \mathsf{E}\Big(\exp\big(\delta(n+1) \;&+\; \varepsilon \sum_{i\geq 0} |X_{n+1-i}| q^i\big) | F_n\Big) \\ &= \exp\left(\delta(n+1) + q\varepsilon \sum_{i\geq 0} |X_{n+1-i}| q^i\right) \mathsf{E}\left(\exp\left(\varepsilon |X_{n+1}|\right) | F_n\right). \end{aligned}$$

Now,

$$\mathsf{E}\left(\exp\left(\varepsilon |X_{n+1}|\right) | F_n\right) \leq \exp\left(\varepsilon \hat{a} \|X^n_{-\infty}\|\right) \mathsf{E} \exp\left(\varepsilon |W_{n+1}|\right),$$

hence, for $n < \tau_m$

$$\begin{aligned} \mathsf{E}\,&(\exp(\delta(n+1) + \varepsilon \|X_{n+1}\|) | F_n) \\ &\leq \exp(\delta(n+1) + \varepsilon(q + \hat{a}) \|X^n_{-\infty}\|) \mathsf{E} \exp(\varepsilon |W_1|) \\ &\leq \exp(\delta n + \varepsilon \|X^n_{-\infty}\|) \exp(\delta) \\ &\quad \times \mathsf{E} \exp(\varepsilon |W_1|) \exp(-\varepsilon(1 - q - a)m). \end{aligned}$$

For any $\delta > 0$ we may take such small $\varepsilon > 0$ and such large $m > 0$ that

$$\mathsf{E} \exp(\varepsilon |W_1|) \exp(-\varepsilon(1 - q - a)m + \delta) \leq \varkappa < 1.$$

Then

$$\begin{aligned} \mathsf{E}\,\mathrm{I}\,&(n < \tau_m) \exp(\delta(n+1) + \varepsilon \|X^{n+1}_{-\infty}\|) \\ &\leq \varkappa \mathsf{E}\,\mathrm{I}\,(n < \tau_m) \exp(\delta(n+1) + \varepsilon \|X^n_{-\infty}\|) \\ &\leq \varkappa \mathsf{E}\,\mathrm{I}\,(n - 1 < \tau_m) \exp(\delta(n+1) + \varepsilon \|X^n_{-\infty}\|). \end{aligned}$$

By induction we obtain

$$\begin{aligned} \mathsf{E}_x \mathrm{I}\,&(n < \tau_m) \exp(\delta(n+1) + \varepsilon \|X^{n+1}_{-\infty}\|) \\ &\leq \varkappa^{n-1} \mathsf{E}_x \mathrm{I}\,(1 < \tau_m) \exp(\delta + \varepsilon \|X^1_{-\infty}\|) \\ &\leq \varkappa^{n-1} \mathsf{E}_x \exp(\delta + \varepsilon \|X^1_{-\infty}\|). \end{aligned}$$

The last expectation possesses the similar bound:

$$\mathsf{E}_x \exp(\delta + \varepsilon\|X_{-\infty}^1\|) \le \exp(\delta + \varepsilon\|x\|)\,\mathsf{E}\exp(\varepsilon|W_1|).$$

thus,

$$\mathsf{E}_x \mathrm{I}(n+1 \le \tau_m)\exp(\delta(n+1) + \varepsilon\|X_{-\infty}^{n+1}\|) \le C(\varepsilon,\delta)\varkappa^{n-1}\exp(\varepsilon\|x\|), \qquad (9.33)$$

where $\varkappa < 1$, $C(\varepsilon,\delta) = \exp(\delta)\,\mathsf{E}\exp(\varepsilon|W_1|)$.

By virtue of (9.33)

$$\begin{aligned}\mathsf{E}_x \exp(\delta\tau) &= \sum_{n\ge 1}\mathsf{E}\,\mathrm{I}(\tau = n)\exp(\delta n) \le \sum_{n\ge 1}\mathsf{E}\,\mathrm{I}(\tau \le n)\exp(\delta n)\\ &\le C(\varepsilon,\delta)\exp(\varepsilon\|x\|)\sum_{n\ge 1}\varkappa^{n-1} = C(\varepsilon,\delta)\exp(\varepsilon\|x\|)(1-\varkappa)^{-1}.\end{aligned}$$

It means that condition (A) holds true with $h_{m,a}^{+}(x) = C(\varepsilon,\delta)\exp(\varepsilon\|x\|)(1-\varkappa)^{-1}$, $h_{m,a}^{-} \equiv 1$.

Conditions (D) and (D_b) may be checked for processes similar to one in example 9.4 if densities of W_n are 'well' enough.

Comments

Results of the chapter are the extension to the non-Markovian case of results of the previous chapter. Theorem 9.2 is almost equivalent to theorem from Orey and Pelikan (1988). Theorem 9.4 is its extension to the 'non-compact' situation. Here 'non-compactness' means the localization of the Doeblin-type condition or its non-Markovian analog. Similar but not equivalent results are obtained with the help of Donsker–Varadhan's and Stroock's approachs in Olla (1987), Chiyonobu and Kusuoka (1988), Deuschel and Stroock (1989) and Nummelin (1990). Nevertheless, recall that in large deviation problems for averaging principle (see Chapter 11 below) it is important that preliminary results such as in Chapters 8 and 9 are obtained with the help of Freidlin's approach.

Considerations are based on the papers Veretennikov (1990) and (1991b).

Chapter 10

Large deviations for empirical measures

10.1. INTRODUCTION

Let X_t be an ergodic process with values in d-dimensional Euclidean space E^d, the time is discrete ($t \in \mathbf{Z}^1 = 0, \pm 1, \pm 2, \ldots$) or continuous ($-\infty < t < +\infty$). Let us define for any Borel set $A \in \mathcal{B}^d$ the random measure

$$\mu_t(A) = t^{-1} \sum_{i=1}^{t} \mathrm{I}(X_i \in A).$$

This formula defines the measure-valued process μ_t with values in the space $\mathsf{M}_1(E^d)$ of all probability measures on E^d. If μ_∞ is an invariant measure then by virtue of the Birkhoff–Khinchin ergodic theorem the convergence $\mu_t \to \mu_\infty$ holds as $t \to \infty$. Denote by Q_t the distribution μ_t on $\mathsf{M}_1(E^d)$. Now we are interested in logarithmic asymptotics of large deviations of Q_t as $t \to \infty$. Such a problem was considered in Stroock (1984), Deuschel and Stroock (1989), Donsker and Varadhan (1975–1983), Varadhan (1984) and Gärtner (1977) for Markov chains and diffusion processes, in Orey and Pelikan (1988) for stationary processes, etc. This chapter is devoted to Markov and stationary cases.

10.2. MARKOV CHAIN WITH DOEBLIN-TYPE CONDITION

Let for any $x, y \in E^d$, $\Gamma \in \mathcal{B}(E^d)$ the following inequality be satisfied:

$$\mathsf{P}(X_1 \in \Gamma | X_0 = x) \leq C\mathsf{P}(X_1 \in \Gamma | X_0 = y). \tag{10.1}$$

Under this condition there exists a stationary measure μ_∞ and $\operatorname{var}(\mu_t - \mu_\infty) \to 0$, $t \to \infty$ (cf. Doob (1953)). Freidlin's method (see Freidlin and Wentzell (1984), Section 9.2) is based on the existence for any $f \in \mathbf{B}(E^d)$ of the limit

$$G(f) = \lim_{t \to \infty} t^{-1} \ln \mathsf{E} \exp(t\langle \mu_t, f \rangle) \tag{10.2}$$

and on some properties of this limit (see below).

Theorem 10.1. Let condition (10.1) be satisfied. Then

(1) for any $f \in \mathbf{B}(E^d)$ the limit (10.2) does exist;
(2) if a bounded sequence (f_n) of Borel functions on E^d converges to $f \in \mathbf{B}(E^d)$ a.e. with respect to measure μ_∞, then $G(f_n) \to G(f)$;
(3) for any $f, g \in \mathbf{B}(E^d)$, $\alpha \in R^1$ the function $G(f + \alpha g)$ is differentiable in α.

Denote by $S(\mu)$ the Legendre transformation of $G(f)$:

$$S(\mu) = \sup_{f \in \mathbf{B}(E^d)} (\langle \mu, f \rangle - G(f)), \qquad \mu \in \mathsf{M}(E^d)$$

($\mathsf{M}(E^d)$ is a set of all σ-additive measures on E^d). Let us define the metric ρ on $\mathsf{M}_1(E^d)$:

$$\rho(\mu, \nu) = \sup (|\langle \mu, f \rangle - \langle \nu, f \rangle|: \ f \in \mathbf{B}(E^d), \|f\|_{\mathrm{B}} \leq 1, \|f\|_{\mathrm{Lip}} \leq 1).$$

Lemma 10.1. Let condition (10.1) be satisfied. Then for any $\delta > 0$ there exist such finite systems $\mathcal{U}$, $\mathcal{V}$ of Borel functions on E^d for which absolute values are bounded by 1 that $|\langle \mu_\infty, w \rangle| \leq \delta$ for $w \in \mathcal{V}$, and for any $f \in \mathbf{B}(E^d)$, $\|f\|_{\mathrm{B}} \leq 1$ there exist such $v \in \mathcal{U}$, $w \in \mathcal{V}$, that $|f - v| \leq w$.

Proof is obvious.

Theorem 10.2. Under conditions of Theorem 10.1:

(1) $S(\mu) < \infty \Rightarrow \mu \geq 0$ and $\mu(E^d) = 1$; functional S is lower semicontinuous in weak convergence of measures topology;
(2) for any $s \geq 0$ the set $(\mu\colon \ S(\mu) \leq s)$ consists on such measures μ that are absolute continuous with respect to μ_∞, and derivatives $\mathrm{d}\mu/\,\mathrm{d}\mu_\infty$ are uniformly bounded on this set;
(3) functional S is lower semicontinuous in topology induced by metrics ρ, and the set $\Phi(s) = (\mu\colon \ S(\mu) \leq s)$, $s < \infty$, is compact in this topology;
(4) let $\mathcal{U}$ be any finite system of functions from $\mathbf{B}(E^d)$,

$$\begin{aligned} \rho_{\mathcal{U}}(\mu, \nu) &= \sup_{v \in \mathcal{U}} |\langle \mu, v \rangle - \langle \nu, v \rangle|, \\ S_{\mathcal{U}}(\mu) &= \sup_{f \in \mathcal{L}(\mathcal{U})} (\langle \mu, f \rangle - G(f)), \end{aligned}$$

where $\mathcal{L}(\mathcal{U})$ is a linear span of functions from $\mathcal{U}$,

$$\Phi_{\mathcal{U}}(s) = (\mu \in \mathsf{M}(E^d)\colon \ S_{\mathcal{U}}(\mu) \leq s), \qquad s \geq 0;$$

then for any $\delta > 0$, $s \geq 0$

$$\inf (S(\nu)\colon \ \rho_{\mathcal{U}}(\nu, \mu) < \delta) \leq S_{\mathcal{U}}(\mu),$$

and

$$\Phi_{\mathcal{U}}(s) \subset (\mu \in \mathsf{M}(E^d)\colon \ \rho_{\mathcal{U}}(\mu, \Phi(s)) < \delta).$$

Theorem 10.3. Let conditions of Theorem 10.1 be satisfied. Then $tS(\mu)$ is the action functional for the family of random measures (μ_t) in metric space $(\mathsf{M}_1(E^d), \rho)$, that is, for any $s > 0$, $\mu \in \mathsf{M}_1(E^d)$ inequalities hold

$$\liminf_{\delta \to 0} \liminf_{t \to \infty} t^{-1} \ln \mathsf{P}(\rho(\mu_t, \mu) < \delta) \geq -S(\mu),$$

$$\limsup_{\delta\to 0}\ \limsup_{t\to\infty}\ t^{-1}\ln \mathsf{P}(\rho(\mu_t, \Phi(s)) \geq \delta) \leq -s.$$

Theorem 10.3 follows from Theorem 9.2.1 in Freidlin and Wentzell (1984) (see Theorem 1.5, Chapter 1 in this book) if assertions of Theorems 10.2 and 10.1 take place.

Proof of Theorem 10.2 follows from considerations of Section 9.2 from Freidlin and Wentzell (1984) and Gärtner (1977).

Proof of Theorem 10.1. I. Let $f \in \mathbf{B}(E^d)$. Denote

$$\begin{aligned} G_n^+(f) &= n^{-1}\ln \sup_x \mathsf{E}_x \exp\langle \mu_n, f\rangle, \\ G_n^-(f) &= n^{-1}\ln \inf_x \mathsf{E}_x \exp\langle \mu_n, f\rangle. \end{aligned}$$

Then like in Stroock (1984) we have for $n, k > 0$

$$G_{n+k}^+(f) \leq \frac{n}{n+k}G_n^+(f) + \frac{k}{n+k}G_k^+(f),$$

$$G_{n+k}^-(f) \geq \frac{n}{n+k}G_n^-(f) + \frac{k}{n+k}G_k^-(f).$$

By virtue of subadditive functions properties (see, for example, Stroock (1984), Lemma 3.11) the following limit exist:

$$G^+(f) = \lim_{n\to\infty} G_n^+(f), \qquad G^-(f) = \lim_{n\to\infty} G_n^-(f).$$

Moreover, by virtue of (10.1)

$$G_n^+(f) \leq n^{-1}\ln C + G_n^-(f). \tag{10.3}$$

Thus, the limit in (10.2) does exist.

II. Compare $G_n^{\pm}(f + \alpha g)$ with definition of spectral radius $r(\alpha)$ of operator T on $\mathbf{B}(E^d)$, where

$$Th(x) = \mathsf{E}_x \exp(\langle \mu_1, f + \alpha g\rangle)\, h(X_1),$$

we see that the equality $G(f + \alpha g) = \ln r(\alpha)$ holds by virtue of (10.1). Function $r(\alpha)$ is differentiable in α (see, for example, Freidlin and Wentzell (1984)), and so is $\ln r(\alpha)$. Property (2) follows from considerations in Gärtner (1977). Theorem 10.1 is proved.

10.3. NON-COMPACT MARKOV CASE

Instead of (10.1) the following stability and non-degeneracy conditions are assumed. There exists such a sequence $(K^m, m \to \infty)$ of bounded Borel sets $K^m \uparrow E^d$ that:

(3A) for any $m \geq 0$ there exist such local bounded functions $h_m^+(x) \geq h_m^-(x) \geq 0$, $x \in E^d$, that $1/h^{\pm}$ are bounded on K^m and for any $a > 0$, $x \in E^d$ for $m \geq m_a$

$$h_m^-(x) \leq \mathsf{E}_x \exp(a\tau_m) \leq h_m^+(x),$$

where $\tau_m := \inf(k \geq 1\colon\ X_k \in K^m)$;

(3B) for any $a > 0$ and $m \geq m_a$, $x \in K^m$, $n \geq 0$

$$C_m^{-1} \leq \mathsf{E}_{0,x} h_m^-(X_n) \leq \mathsf{E}_{0,x} h_m^+(X_n) \leq C_m, \qquad C_m > 0,$$

and $\lim_{m\to\infty} C_m = 1$;

(3C) $\lim\limits_{m\to\infty} \inf\limits_{x\in K^m} \mathsf{P}_{0,x}(X_1 \in K^m) = 1$;

(3D) for any $m \geq m_0$

$$\inf_{x,y\in K^m} \int_{K^m} 1 \wedge \frac{\mathsf{P}(y, 1, dz)}{\mathsf{P}(x, 1, dz)} \mathsf{P}(x, 1, dz) > 0,$$

where $d\mathsf{P}'/d\mathsf{P}$ means the density of the absolute continuous component of P' w.r.t. P;

(3E) process $(X_{\tau_m^n}, n \geq 0)$ for any $m \geq m_0$ has a stationary ergodic distribution μ_m on K^m,

$$\sup_x \operatorname{var}(\mathcal{L}(X_{\tau_m^n} | X_0 = x) - \mu_m) \to 0, \qquad n \to \infty,$$

where $\tau_m^1 = \tau_m$, $\tau_m^{i+1} = \theta_{\tau_m} \tau_m^i$, $i \geq 1$;

(3F) for any $V \in \mathcal{B}(E^d)$ the function $\mathsf{P}_y(X_1 \in V \cap K^m)$ is continuous in y for any m.

Condition (3D) is a local Doeblin-type condition.

Under conditions (3A)–(3F) the unique stationary measure μ_∞ exists (see Nummelin (1984)).

Theorem 10.4. Let conditions (3A)–(3F) be satisfied. Then

(1) the limit (10.2) exists for any $f \in \mathbf{B}(E^d)$;
(2) if bounded sequence (f_n) of measurable functions converges to $f \in \mathbf{B}(E^d)$ a.e. w.r.t. μ_∞, then $G(f_n) \to G(f)$.

Note that property (3) of Theorem 10.1 is not asserted here. Functional $S(\mu)$ and metric ρ on $\mathsf{M}_1(E^d)$ are introduced in the same way as earlier.

Theorem 10.5. Let conditions of Theorem 10.4 be satisfied. Then assertions (1)–(4) of Theorem 10.2 hold for the process (X_t).

Theorem 10.6. Let conditions of Theorem 10.4 be satisfied. Then $tS(\mu)$ is an action functional for the family of random measures (μ_t) in metric space $(\mathsf{M}_1(E^d), \rho)$, that is, for any $s > 0$, $\mu \in \mathsf{M}_1(E^d)$ inequalities hold

$$\liminf_{\delta\to0} \liminf_{t\to\infty} t^{-1}\mathsf{P}(\rho(\mu_t, \mu) < \delta) \geq -S(\mu),$$

$$\limsup_{\delta\to0} \limsup_{t\to\infty} t^{-1}\mathsf{P}(\rho(\mu_t, \Phi(s)) \geq \delta) \leq -s.$$

Note that differentiability of function $G(f + \alpha g)$ in α is not stated so Theorem 10.6 is not so trivial as it does not follow directly from previous two theorems as Theorem 10.3.

Proof of Theorem 10.4 follows from considerations analogous to those in the proof of Theorem 1.1 from Veretennikov (1990) (where the diffusion process was studied).

Proof of Theorem 10.5 is just like the proof of Theorem 10.2.

Proof of Theorem 10.6. I. The second inequality follows from considerations in Gärtner (1977), see also Theorem 9.2.1 from Freidlin and Wentzell (1984). In this part we do not need the differentiability.

II. The first inequality is reduced to the finite-dimensional case with the help of Theorem 10.5 and the inequality

$$\mathsf{P}(\rho(\mu_t,\mu) < \delta) \geq \mathsf{P}(\rho_{\mathcal{U}_\delta}(\mu_t,\mu) < \delta/2) - \mathsf{P}\left(\max_{w\in\mathcal{V}_\delta}\langle\mu_t, w\rangle \geq \delta/4\right)$$

(see Freidlin and Wentzell (1984), Section 9.2 and Theorem 1.5, Chapter 1 in this book).

III. The proof of lower estimate in the finite-dimensional case without differentiability of G may be found in Chapter 8. Theorem 10.6 is proved.

10.4. STATIONARY COMPACT CASE

Consider stationary process X_t with discrete time, $t \in \mathbf{Z}^1$. Denote $X_b^a = (\mathbf{X}_t, a \leq t \leq b)$. Assume that for any $N > 0$, $\Gamma \in \mathcal{B}^N$ and any $x = (x_0, x_{-1}, \ldots)$, $y = (y_0, y_{-1}, \ldots)$

$$\mathsf{P}(X_1^N \in \Gamma | X_{-\infty}^0 = x) \leq C_N \mathsf{P}(X_1^N \in \Gamma | X_{-\infty}^0 = y), \tag{10.4}$$

$$\lim_{N\to\infty} N^{-1} \ln C_N = 0.$$

This is a non-Markov analogy of condition (10.1), (10.4) is a bit stronger than $(R\text{–}M)$ (ratio mixing), see Orey and Pelikan (1988). We may also use the $(R\text{–}M)$ condition but in the next section we consider the more general case and more general 'non-compact' condition.

Theorem 10.7. Let condition (10.4) be satisfied. Then

(1) for any $f \in \mathbf{B}(E^d)$ the limit (2.2) exists;
(2) if bounded sequence (f_n) of Borel functions on E^d converge to $f \in \mathbf{B}(E^d)$ a.e (μ_∞), then

$$G(f_n) \to G(f).$$

Note that differentiability of G is not asserted. As previous, functional S and metric ρ are introduced.

Theorem 10.8. Let conditions of Theorem 10.7 be satisfied. Then assertions (1)–(4) of Theorem 10.2 hold for (X_t).

Theorem 10.9. Let conditions of Theorem 10.7 be satisfied. Then $tS(\mu)$ is an action functional for the family of random measures (μ_t) in metric space $(\mathsf{M}_1(E^d), \rho)$, i.e., for any $s > 0$, $\mu \in \mathsf{M}_1(E^d)$

$$\liminf_{\delta\to 0}\ \liminf_{t\to\infty}\ t^{-1}\ln \mathsf{P}(\rho(\mu_t,\mu) < \delta) \geq -S(\mu),$$

$$\limsup_{\delta\to 0}\ \limsup_{t\to\infty}\ t^{-1}\ln \mathsf{P}(\rho(\mu_t,\Phi(s)) \geq \delta) \leq -s.$$

Proof of Theorem 10.7 follows from considerations in the first step of the proof of Theorem 2.1. By virtue of (10.4) we obtain an inequality

$$G_n^+(f) \leq n^{-1}\ln C_n + G_n^-(f).$$

instead of (10.3). By virtue of the second part of (10.4) this implies the equality $G^+(f) = G^-(f)$, and so the existence of the limit in (10.2) is stated. Property (2) also follows from considerations in Gärtner (1977). Theorem 10.7 is proved.

Proof of Theorem 10.8 is analogous to the proof of Theorem 10.2.

Proof of Theorem 10.9 is analogous to the proof of Theorem 3.3. Since it is the non-Markov case then we must use results from Chapter 9 instead of Chapter 8 in step III. Theorem 10.9 is proved.

10.5. STATIONARY NON-COMPACT CASE

Consider the process X_t, $t \in \mathbf{Z}^1$. Denote $Z_- = (0,-1,-2,\ldots)$, $Z_+ = (1,2,\ldots)$. Assume that the following stability and non-degeneracy conditions hold: there exist such sequence $(K^m, m \geq 0)$ of Borel sets in $(E^d)^{Z_-}$ that $(K^m \uparrow E^d)^{Z_-}$ and

(5A) for any $m \geq 0$ there exist such finite functions $h_m^+(x) \geq h_m^-(x) > 0$, $\mathbf{x} = (x_0, x_{-1}, \ldots) \in (E^d)^{Z_-}$, that for any $a > 0$, $m \geq m_0$, $x \in K^m$

$$h_m^-(x) \leq \mathsf{E}(\exp(a\tau_m)|X_0^{-\infty} = x) \leq h_m^+(x),$$

where $\tau_m = \inf(k \geq 1\colon X_k^{-\infty} \in K^m)$ we always may choose $(h_m^- \equiv 1)$;

(5B) for any $a > 0$, $m \geq m_a$, $n \geq 0$, $x \in (E^d)^{Z_-}$

$$C_m^{-1} \leq \mathsf{E}(h_m^-(X_n^{-\infty})|X_0^{-\infty} = x) \leq \mathsf{E}(h_m^+(X_n^{-\infty})|X_0^{-\infty} = x) \leq C_m, \qquad C_m > 0,$$

and $\lim_{m\to\infty} C_m = 1$, where $h^\pm$ are functions from (5A);

(5C) $\displaystyle\liminf_{m\to\infty}\ \inf_{x\in K^m} \mathsf{P}(X_1^{-\infty} \in K^m | X_0^{-\infty} = x) = 1$;

(5D) for any $m \geq 0$ there exist such constants $L_m(N)$, $N \geq 1$, that

$$\lim_{N\to\infty} N^{-1}\ln L_m(N) = 0$$

and for any $x, y \in K^m$ and $\Gamma_N \in \mathcal{B}((E^d)^N)$

$$\mathsf{P}(X_N^1 \in \Gamma | X_0^{-\infty} = x) \leq L_m(N)\,\mathsf{P}(X_N^1 \in \Gamma | X_0^{-\infty} = y).$$

These conditions are sufficient for the existence of stationary distribution.

Theorem 10.10. Let conditions (5A)–(5D) be satisfied. Then for any $f \in \mathbf{B}(E^d)$ the limit (10.2) does exist, and if bounded sequence (f_n) of Borel functions on E^d converges to $f \in \mathbf{B}(E^d)$ a.e. (μ_∞), then $G(f_n) \to G(f)$.

S and ρ are introduced as earlier.

Theorem 10.11. Let conditions of Theorem 10.10 be satisfied. Then assertions (1)–(4) of Theorem 2.2 hold for (X_t).

Theorem 10.12. Let conditions of Theorem 10.10 be satisfied. Then $tS(\mu)$ is an action functional for the family of random measures (μ_t) in $(\mathsf{M}_1(E^d), \rho)$, i.e., for any $s > 0$, $\mu \in \mathsf{M}_1(E^d)$ the following inequalities hold

$$\liminf_{\delta\to 0} \liminf_{t\to\infty} t^{-1} \ln \mathsf{P}(\rho(\mu_t, \mu) < \delta) \geq -S(\mu),$$

$$\limsup_{\delta\to 0} \limsup_{t\to\infty} t^{-1} \ln \mathsf{P}(\rho(\mu_t, \Phi(s)) \geq \delta) \leq -s.$$

Proofs of Theorems 10.10–10.12 are analogous to proofs in previous sections. In Theorem 10.12 one must use results of Chapter 9 after reduction to the finite-dimensional case.

Comments

In this chapter we extend to 'level 2' the results of Chapters 8 and 9. This extension is similar to Gärtner's.

Chapter 11

Large deviations in averaging principle

We now go to our main aim, that is, the study of large deviations in averaging principle. We firstly consider the 'compact' case in methodical reasons as in previous chapters. We study here the Markovian case only.

11.1. COMPACT CASE

Let us consider large deviations in averaging principle for a stochastic difference system

$$\begin{aligned} \vartheta_{n+1} &= \vartheta_n + \gamma f(\vartheta_n, Y_n), \\ Y_{n+1} &= Y_n + a(\vartheta_n, Y_n) + b(\vartheta_n, Y_n)W_n, \\ \vartheta_0 &= \vartheta, \qquad Y_0 = y, \end{aligned} \tag{11.1}$$

on the torus $T^d \times T^l$. Here $\vartheta_n \in T^d$, $Y_n \in T^l$ ($T^d = [0,1)^d$ is the torus of dimension d), f is a function with values in d-dimensional Euclidean space E^d, a is a function with values in E^l, b is a $l \times l$ matrix, $W_n \in E^l$, $\gamma > 0$ is a small parameter, $(W_n, n = 0, 1, \dots)$ is a sequence of i.i.d. random variables on some probability space (Ω, F, P). These equations are understood as follows: if $\vartheta_n \in T^d$, $Y_n \in T^l$ are done then we must calculate right-hand sides of equations (11.1) and then apply the function $\{\cdot\}(\{a\} = a - [a]$, where $[a]$ is the integer part of the value a).

Assume the following condition holds:

A. Functions f, a, b are bounded, the matrix b is uniformly non-degenerate:

$$\inf_{\vartheta, y} \inf_{|\lambda|=1} \lambda^* b b^*(\vartheta, y)\lambda > 0;$$

functions f, a, b are continuously differentiable in all variables; the distribution of W_1 has a continuous positive density.

Let us consider the limit

$$G(\vartheta, z, \alpha) = \lim_n n^{-1} \ln \mathsf{E} \exp\left(\alpha \sum_{k=1}^{n} f(\vartheta, y_k^z)\right), \tag{11.2}$$

where the process y^z satisfies the equation (in the same sense as (11.1))

$$y_{n+1}^z = y_n^z + a(z, y_n^z) + b(z, y_n^z)W_n, \qquad y_0^z = y.$$

If the limit in (11.2) exists then similar to the continuous time case one may hope that the asymptotics of the process ϑ_n, $0 \le n \le T/\gamma$, as $\gamma \to 0$ is described with the help of the action functional

$$S_{0T}(\varphi) = \int_0^T L(\varphi_s, \dot\varphi_s)\,\mathrm{d}s, \tag{11.3}$$

where $L(\vartheta,\beta)$ is a Fenchel–Legendre transformation of the function $G(\vartheta,\vartheta,\alpha)$:

$$L(\vartheta,\beta) = \sup(\alpha\beta - G(\vartheta,\vartheta,\alpha)).$$

Fix the value $T > 0$ and denote

$$\tilde{\vartheta}^{\gamma}_s = \sum_k \vartheta_k \mathrm{I}([s/\gamma] \le k < [s/\gamma] + 1).$$

Theorem 11.1. Let condition A be satisfied. Then for any α, ϑ, z, y there exists the limit (11.2) uniformly in ϑ, z, y and $|\alpha| \le a$ for any finite $a > 0$ function G does not depend on y, G is continuous in all variables, convex in α and differentiable in α.

Theorem 11.2. Let condition A be satisfied. Then $\gamma^{-1}S_{0T}(\varphi)$ is an action functional for the family $(\tilde{\vartheta}^{\gamma}_t, t \le T)$ in $C[0,T;T^d]$, that is, for any $x \in T^d$, $s > 0$ the set $\Phi_\vartheta(s) = (\varphi\colon \varphi \in C[0,T;T^d],\ \varphi_0 = \vartheta,\ S_{0T}(\varphi) \le s)$ is a compact in $C[0,T;T^d]$ and for any $s > 0$, $\delta > 0$ the following inequalities hold:

$$\liminf_{\delta\to 0}\ \liminf_{\gamma\to 0}\ \gamma \ln \mathrm{P}(\rho_{0T}(\tilde{\vartheta}^{\gamma},\varphi) < \delta) \ge -S_{0T}(\varphi), \tag{11.4}$$

$$\limsup_{\delta\to 0}\ \limsup_{\gamma\to 0}\ \gamma \ln \mathrm{P}(\rho_{0T}(\tilde{\vartheta}^{\gamma},\Phi_\vartheta(s)) < \delta) \le -s. \tag{11.5}$$

Let μ^z be the invariant measure for the process y^z. Denote

$$\bar{f}(\vartheta) = \int f(\vartheta,y)\,\mu^{\vartheta}(\mathrm{d}y).$$

Consider the equation

$$\dot{\bar{\vartheta}}_t = \bar{f}(\bar{\vartheta}_t), \qquad \bar{\vartheta}_0 = \vartheta. \tag{11.6}$$

Functional S equals zero on any solution of (11.6) and $S \ne 0$ on any other path (see Freidlin (1978) and Freidlin and Wentzell (1984)).

Proof of Theorem 11.1. The main assertion about the existence and differentiability of the limit follows from the same considerations as in Freidlin (1978) and Freidlin and Wentzell (1984), that is, from Frobenius-type theorem on a spectral radius of a positive operator.

The boundedness of function f and equality (11.2) imply the Lipschitz condition for function G in α with a unique constant for any ϑ, z.

By virtue of (11.2) and uniform continuity of f in ϑ function G is uniformly continuous in ϑ.

(Uniform) continuity of G in z will be also proved if we show that the limit in (11.2) is uniform in any ϑ, z, and any $|\alpha| \le a$ for any a fixed. It is sufficient for this aim to prove that eigen-functions that correspond the spectral radii of operators $T^{\alpha,\vartheta,z}$ on $B(T^l)$ given by the formula

$$T^{\alpha,\vartheta,z}h(y) = \mathrm{E}_y h(y^z_1)\exp(\alpha f(\vartheta,y))$$

are uniformly bounded and uniformly separated from zero (cf. considerations in the proof of Theorem 7.2.4 in Freidlin and Wentzell (1984)). However, the latter property is a simple consequence of the equation for the eigen-function $Th(y) = \lambda h(y)$, condition A and the fact that all spectral radii $r(T^{\alpha,\vartheta,z})$ are uniformly bounded and uniformly separated from zero for $|\alpha| \le a$. The latter assertion follows, for example, from the well known bound for a spectral radius of a positive operator (see Krasnoselskii, Lifshitz and Sobolev (1989)): if $Th \ge \alpha h$ for some $\alpha \ge 0$, $h > 0$, then $r(T) \ge \alpha$. In our case we may take $h = 1$ and use the inequality $\inf_{|\alpha|\le a} \inf_{\vartheta,z,y} T^{\alpha,\vartheta,z}\mathbf{1}(y) > 0$ which follows from conditions. Theorem 11.1 is proved.

Proof of Theorem 11.2. I. Since function f is bounded then G grows in α not faster than some linear function, so L is equal to $+\infty$ outside some bounded set of variables β. Hence, including $\varphi \in \Phi_x(s)$ implies the inequality ess sup $|\dot{f}_t| < \infty$. Thus, to prove the compactness of the set $\Phi_x(s)$ it is sufficient to establish the lower semicontinuity of functional S_{0T}. However, this is proved by the same considerations as in Freidlin (1978) and Freidlin and Wentzell (1984).

II. Let us choose such (small) $\Delta > 0$ that $T/\Delta = m$ is an integer. Let $\psi_t = \varphi_{[t/\Delta]\Delta}$, $\lambda > 0$, χ_t is such a piecewise-linear function with values in T^d which derivative jumps in point $k\Delta$ that $\rho_{0T}(\varphi,\chi) < \lambda$, where $\rho_{0T}(\varphi,\chi) = \sup_{0\le t\le T} |\varphi_t - \chi_t|$. Assume Δ/γ to be integer for simplicity.

Let us denote by y^ψ, Y^ψ, X^ψ, $\tilde{X}^\psi$ the processes

$$y^\psi_{n+1} = y^\psi_n + a(\psi_{n\gamma}, y^\psi_n) + b(\psi_{n\gamma}, y^\psi_n)W_n, \qquad y^\psi_0 = y,$$

$$X^\psi_{k\Delta} = \tilde{X}^\psi = \tilde{\vartheta}^\gamma_{k\Delta}, \qquad X^\psi_{(n+1)\gamma} = X^\psi_{n\gamma} + \gamma f(\psi_{n\gamma}, Y^\psi_{n\gamma}), \qquad Y^\psi_{n\gamma} = y^\psi_n,$$

$$\tilde{X}^\psi_{(n+1)\gamma} = \tilde{X}^\psi_{n\gamma} + \gamma f(\psi_{n\gamma}, Y_n), \qquad k\Delta \le n\gamma < (k+1)\Delta, \quad k = 0,\dots,m-1.$$

Denote

$$\tilde{S}_{0T}(\varphi) = \tilde{S}^\psi_{0T}(\varphi) = \int_0^T L(\psi_s, \dot{\varphi}_s)\,ds.$$

This functional is lower semicontinuous in (ψ,φ) in uniform topology. This is proved precisely as in Freidlin (1978) and Freidlin and Wentzell (1984).

Let us denote $T_k = k\Delta$, $F_t = \sigma(W_k\colon k\gamma \le t)$. Show that for any $\nu > 0$, $\Delta > 0$ small enough and $\gamma \le \gamma(\nu,\Delta)$ the following inequality holds:

$$\mathsf{P}(\rho_{0T}(\tilde{\vartheta}^\gamma, \varphi) < \delta) \ge \exp(-\gamma^{-1}(\tilde{S}^\psi_{0T}(\varphi) + \nu)). \tag{11.7}$$

To this aim we use another method of existence of the limit (11.2) proof which is based on subadditive functions properties. Let us denote

$$G^+_n(x,z,\alpha) = \sup_y n^{-1} \ln \mathsf{E}_y \exp\left(\alpha \sum_{k=1}^n f(x, y^z_k)\right),$$

$$G^-_n(x,z,\alpha) = \inf_y n^{-1} \ln \mathsf{E}_y \exp\left(\alpha \sum_{k=1}^n f(x, y^z_k)\right).$$

With the help of the Markovian property we have for $k \leq n$

$$G_n^+(x,z,\alpha) \leq (k/n)G_k^+(x,z,\alpha) + (n-k/n)G_{n-k}^+(x,z,\alpha),$$

which implies by virtue of subadditive functions (see, for example, Stroock (1984)) the existence of the limit

$$G^+(x,z,\alpha) = \lim_n G_n^+(x,z,\alpha) = \inf_n G_n^+(x,z,\alpha).$$

Similarly for G_n^- we obtain the existence of the limit

$$G^-(x,z,\alpha) = \lim_n G_n^-(x,z,\alpha) = \sup_n G_n^-(x,z,\alpha).$$

At last, by virtue of compactness of the torus and non-degeneracy and boundedness of the density of W_n we obtain for any n and any y, y'

$$\mathsf{E}_y \exp\left(\alpha \sum_{k=1}^n f(x, y_k^z)\right) \leq C\mathsf{E}_{y'} \exp\left(\alpha \sum_{k=1}^n f(x, y_k^z)\right),$$

which implies the equality

$$G^-(x,z,\alpha) = G^+(x,z,\alpha),$$

and the existence of the limit uniformly in all x and z

$$G(x,z,\alpha) = \lim_n G_n(x,z,\alpha)(= G_n^\pm(x,z,\alpha)).$$

Since the limit is uniform then for any $\nu > 0$ there exists such N that

$$\sup |G(x,z,\alpha) - G_N(x,z,\alpha)| \leq \nu/2.$$

Function $G_N(x,z,\alpha)$ is uniformly continuous in z for $|\alpha| \leq a$ and N fixed. Hence, there exists such $\varepsilon > 0$ that for

$$|z'-z| + |x-x'| + |\alpha - \alpha'| < \varepsilon, \qquad |\alpha|, |\alpha'| \leq a$$

$$|G_N(x',z',\alpha') - G_N(x,z,\alpha)| \leq \nu/2.$$

Thus,

$$|G(x',z',\alpha') - G(x,z,\alpha)| \leq \nu.$$

III. Now, let us denote for $z = (z.)$, $x = (x.)$

$$y_{n+1}^z = y_n^z + a(z_n, y_n^z) + b(z_n, y_n^z)W_n, \qquad y_0^z = y,$$

$$G_n^+(x,z,\alpha;\varepsilon) = \sup_y \sup_{\substack{|z_k - z| < \varepsilon,\\ |x_k - x| < \varepsilon,\\ |\alpha_k - \alpha| < \varepsilon, \quad k \leq n}} n^{-1} \ln \mathsf{E}_y \exp\left(\sum_{k=1}^n \alpha_k f(x., y_k^z)\right),$$

$$G_n^-(x,z,\alpha;\varepsilon) = \inf_y \inf_{\substack{|z_k - z| < \varepsilon,\\ |x_k - x| < \varepsilon,\\ |\alpha_k - \alpha| < \varepsilon, \quad k \leq n}} n^{-1} \ln \mathsf{E}_y \exp\left(\sum_{k=1}^n \alpha_k f(x., y_k^z)\right).$$

By virtue of the Markovian property and continuity we have

$$G_n^+(x,z,\alpha;\varepsilon) \le \ln C_\varepsilon + G_n^+(x,z,\alpha),$$

where $C_\varepsilon \to 1$ as $\varepsilon \to 0$. Similarly

$$G_n^-(x,z,\alpha;\varepsilon) \ge -\ln C_\varepsilon + G_n^+(x,z,\alpha)$$

with the same property for C_ε. This implies that for $|z'-z|+|x-x'|+|\alpha-\alpha'|<\varepsilon$, $|\alpha|$, $|\alpha'| \le a$

$$\begin{aligned} G(x,z,\alpha) - \ln C_\varepsilon &\le \liminf_{n\to\infty} G_n^-(x',z',\alpha';\varepsilon) \le \limsup_{n\to\infty} G_n^+(x',z',\alpha';\varepsilon) \\ &\le G(x,z,\alpha) + \ln C_\varepsilon. \end{aligned} \tag{11.8}$$

IV. Let us consider the following Cramér-type measure transformations on F_{T_j}:

$$d\mathsf{P}^{\gamma,\alpha}/\,d\mathsf{P} = \exp(\gamma^{-1}(\alpha(X_T^\psi - X_{T_{m-1}}^\psi) - \hat h_\Delta^\gamma)),$$

$$d\tilde{\mathsf{P}}^{\gamma,\alpha}/\,d\mathsf{P} = \exp(\gamma^{-1}(\alpha(\tilde\vartheta_T^\gamma - \tilde\vartheta_{T_{m-1}}^\gamma) - \tilde h_\Delta^\gamma)),$$

where

$$\begin{aligned} \hat h_\Delta^\gamma &= \gamma \ln \mathsf{E}_y(\exp(\gamma^{-1}\alpha(X_T^\psi - X_{T_{m-1}}^\psi))|F_{T_{m-1}}), \\ \tilde h_\Delta^\gamma &= \gamma \ln \mathsf{E}_y(\exp(\gamma^{-1}\alpha(\tilde\vartheta_T^\gamma - \tilde\vartheta_{T_{m-1}}^\gamma))|F_{T_{m-1}}). \end{aligned}$$

Measure $\mathsf{P}^{\gamma,\alpha}$, $\tilde{\mathsf{P}}^{\gamma,\alpha}$ are probabilistic ones. By virtue of (11.8) for any $\nu > 0$ there exist such $\Delta > 0$, $\lambda > 0$, $\delta' > 0$ that on the set $(\rho_{0,T_{m-1}}(\tilde\vartheta^\gamma,\varphi) < \delta')$

$$\begin{aligned} \Delta(G(\psi_{T_{m-1}},\psi_{T_{m-1}},\alpha) - \nu) &\le \liminf_{\gamma\to 0}\ h_\Delta^\gamma \le \limsup_{\gamma\to 0}\ h_\Delta^\gamma \\ &\le \Delta(G(\psi_{T_{m-1}},\psi_{T_{m-1}},\alpha) + \nu). \end{aligned} \tag{11.9}$$

Then we have on the set $(\rho_{0,T_{m-1}}(\tilde\vartheta^\gamma,\varphi) < \delta')$ for δ'' small enough

$$\begin{aligned} \mathsf{P}(\rho_{T_{m-1},T}(\tilde\vartheta^\gamma,\varphi) < \delta | F_{T_{m-1}}) &\ge \mathsf{E}\{\mathrm{I}(\rho(\tilde\vartheta_T^\gamma,\varphi_T) < \delta'')|F_{T_{m-1}}\} \\ &= \tilde{\mathsf{E}}^{\gamma,\alpha}\{\mathrm{I}(\rho(\tilde\vartheta_T^\gamma,\varphi_T) < \delta'')(\,d\tilde{\mathsf{P}}^{\gamma,\alpha}/\,d\mathsf{P})^{-1}|F_{T_{m-1}}\} \\ &\ge \mathsf{E}^{\gamma,\alpha}\{\mathrm{I}(\rho(\tilde\vartheta_T^\gamma,\varphi_T) < \delta'') \\ &\quad \times \exp(-\gamma^{-1}\nu - \gamma^{-1}\Delta[\alpha\Delta^{-1}(\chi_T - \chi_{T_{m-1}}) \\ &\quad -G(\psi_{T_{m-1}},\psi_{T_{m-1}},\alpha)]|F_{T_{m-1}}\}. \end{aligned} \tag{11.10}$$

Now choose such $\alpha = \alpha_{m-1}$ that

$$\begin{aligned} L(\psi_{T_{m-1}},\beta) &> L(\psi_{T_{m-1}},\Delta^{-1}(\chi_T - \chi_{T_{m-1}})) \\ &\quad +\alpha_{m-1}(\beta - \Delta^{-1}(\chi_T - \chi_{T_{m-1}})), \end{aligned}$$

$$\beta \neq \Delta^{-1}(\chi_T - \chi_{T_{m-1}}).$$

It is possible by virtue of strict convexity of function L in β. Moreover,

$$\begin{aligned} G(\psi_{T_{m-1}}, \psi_{T_{m-1}}, \alpha_{m-1}) &= \alpha_{m-1}(\Delta^{-1}(\chi_T - \chi_{T_{m-1}})) \\ &\quad - L(\psi_{T_{m-1}}, \Delta^{-1}(\chi_T - \chi_{T_{m-1}})). \end{aligned}$$

Thus,

$$\begin{aligned} &\mathsf{P}(|\tilde{\vartheta}_T^\gamma - \varphi_T| < \delta'' | F_{T_{m-1}}) \\ &\quad \geq \exp(-\gamma^{-1}\nu/2 - \gamma^{-1}\Delta L(\psi_{T_{m-1}}, \Delta^{-1}(\chi_T - \chi_{T_{m-1}}))) \\ &\quad \times \mathsf{E}^{\gamma,\alpha_{m-1}} \mathrm{I}(|\tilde{\vartheta}_T^\gamma - \varphi_T| < \delta'' | F_{T_{m-1}}). \end{aligned}$$

By virtue of Theorem 5.1.1 from Freidlin and Wentzell (1984) applied to the family of measures $\mathsf{P}^{\gamma,\alpha_{m-1}}$

$$\mathsf{P}^{\gamma,\alpha_{m-1}}(|\tilde{\vartheta}_T^\gamma - \varphi_T| < \delta'' | F_{T_{m-1}}) \to 1, \qquad \gamma \to 0,$$

that checked by the same considerations as in the proof of Theorem 5.1.2 from Freidlin and Wentzell (1984). Item, for γ small enough we obtain

$$\begin{aligned} &\mathsf{E}\{\mathrm{I}(\rho_{T_{m-1},T}(\tilde{\vartheta}^\gamma, \varphi) < \delta) | F_{T_{m-1}}\} \\ &\qquad \geq \exp(-\gamma^{-1}(\Delta(L(\psi_{T_{m-1}}, \Delta^{-1}(\chi_T - \chi_{T_{m-1}}))) + \nu)). \end{aligned}$$

By induction changing δ' and δ'' every time if necessary we have

$$\begin{aligned} &\mathsf{E}_{\vartheta,x,y}\mathrm{I}(\rho_{0T}(\tilde{\vartheta}^\gamma, \varphi) < \delta) \\ &\qquad \geq \exp\left(-\gamma^{-1}\sum_{k=1}^{m} \Delta L(\psi_{T_{k-1}}, \Delta^{-1}(\chi_{T_k} - \chi_{T_{k-1}})) - \gamma^{-1}\nu\right). \end{aligned}$$

V. The following equality holds:

$$\sum_{k=1}^{m} \Delta L(\psi_{T_{k-1}}, \Delta^{-1}(\chi_{T_k} - \chi_{T_{k-1}})) = \int_0^T L(\psi_s, \dot{\varphi}_s)\,\mathrm{d}s.$$

Then we have as $\gamma \to 0$

$$\mathsf{P}(\rho_{0T}(\tilde{\vartheta}^\gamma, \varphi) < \delta) \geq \exp\left(-\gamma^{-1}\left(\int_0^T L(\psi_s, \dot{\varphi}_s)\,\mathrm{d}s + \nu\right)\right).$$

By virtue of convexity of L in the second variable β this inequality implies, similar to Freidlin (1978) and Freidlin and Wentzell (1984),

$$\mathsf{P}(\rho_{0T}(\tilde{\vartheta}^\gamma, \varphi) < \delta) \geq \exp(-\gamma^{-1}(\tilde{S}_{0T}^\psi(\chi) + \nu)).$$

VI. Finally, by virtue of Lemma 7.5.2 from Freidlin and Wentzell (1984) for any sequence of step functions $\psi^{(n)} \to \varphi \in C[0,T;T^d]$ there exists such sequence of piecewise-linear functions $\varphi^{(n)} \in C[0,T;T^d]$ that

$$\limsup_{n\to\infty} \int_0^T L(\psi_s^{(n)}, \dot{\varphi}_s^{(n)})\,\mathrm{d}s \leq S_{0T}(\varphi).$$

Thus, by virtue of properties of function L we have as $\gamma \to 0$

$$\mathsf{P}(\rho_{0,T}(X^{\psi^\lambda}, \varphi^\lambda) < \delta) \geq \exp(-\gamma^{-1}(S_{0T}(\varphi) + \nu)),$$

which coincides with (11.4).

VII. Proof of the second inequality (11.5) follows from considerations in the proof of the second inequality in Theorem 7.4.1 from Freidlin and Wentzell (1984) without changes. Theorem 11.2 is proved.

11.2. NON-COMPACT CASE

Assume conditions A, B and C from Chapter 7 are fulfilled. We recall them for the convenience of the reader:

A. The matrix $b(\vartheta, y)$ is uniformly non-degenerate, i.i.d. random values W_n have a positive density w.r.t. Lebesgue measure, the vector $a(\vartheta, y)$ and the matrix $b(\vartheta, y)$ are continuous in ϑ uniformly w.r.t. y.

B. For any $c > 0$ there exist such $R_0 > 0$, $\nu > 0$ that for any x, z, y, $R \geq R_0$

$$\mathsf{E}_y \exp(c\tau_z^R) \leq \exp(\nu(|y| - R)_+),$$

where $\tau^R = \tau_z^R = \inf(t \geq 0\colon\ |y_t^z| \leq R)$.

C. For any $K > 0$ and $\nu > 0$ there exists such $R_0 > 0$ that for any $R \geq R_0$ and s, x, y, z

$$\mathsf{E}_y \exp(\nu(|y_s^z| - R)_+) \leq K \exp(\nu(|y| - R)_+),$$

and for any $\gamma \leq \gamma_0$

$$\mathsf{E}_y \exp(\nu(|Y_s| - R)_+) \leq K \exp(\nu(|y| - R)_+).$$

Let us introduce the process y_n^z as earlier, that is,

$$y_{n+1}^z = y_n^z + a(z, y_n^z) + b(z, y_n^z)W_n, \qquad y_0^z = Y_0.$$

Consider the limit

$$G(\vartheta, \alpha) = \lim_{n\to\infty} n^{-1} \ln \mathsf{E} \exp\left(\alpha \sum_{k=1}^{n} f(\vartheta, y_n^\vartheta)\right). \tag{11.11}$$

If the limit exists then as in 11.1 one may hope that the action functional for ϑ_n, $0 \leq n \leq T/\gamma$, as $\gamma \to 0$ is given by the expression

$$S_{0T}(\varphi) = \int_0^T L(\varphi_s, \dot\varphi_s)\, ds \tag{11.12}$$

where $L(\vartheta, \beta)$ is the Fenchel–Legendre transformation of function

$$G\colon\ L(\vartheta, \beta) = \sup(\alpha\beta - G(\vartheta, \alpha)).$$

This is proved below.

Theorem 11.3. Let conditions A, B and C from Chapter 7 be satisfied, $f \in \mathbf{B}(E^d)$, $f(x) \in E^l$. Then for any ϑ, α the limit (11.11) exists, it is finite, does not depend on y, is convex in α and continuous in (ϑ, α). For any $\alpha \in E^l$, $\lambda > 0$, $\delta > 0$, $\vartheta > 0$ there exist such constants n_0, m, $C(m)$, m that for any $|\vartheta| \le \vartheta$, Y_0, and $n \ge n_0$

$$\begin{aligned} -\delta - C(m)n^{-1} &+ n^{-1}\ln\max\{\exp(-K_\vartheta n);\ 1-\exp(-\lambda n)\,h_m(\vartheta_0)\} \\ &\le n^{-1}\ln \mathsf{E}_y \exp\left(\sum_{k=1}^{n} \alpha f(\vartheta, y_k^\vartheta)\right) - G(\vartheta,\alpha) \\ &\le \delta + C(m)n^{-1} \\ &\quad + n^{-1}\ln\min\{\exp(K_\vartheta n);\ 1+\exp(-\lambda n)h_m(\vartheta_0)\} \end{aligned} \tag{11.13}$$

(here $K_\vartheta = \|f(\vartheta,\cdot)\|_{\mathbf{B}}$).

Let us denote

$$L(\vartheta,\beta) = \sup(\alpha\beta - G(\vartheta,\alpha))$$

and

$$S_{0T}(\varphi) = \int_0^T L(\varphi_s, \dot\varphi_s)\,\mathrm{d}s.$$

Theorem 11.4. Let conditions of Theorem 11.3 be satisfied. Then $\gamma^{-1}S_{0T}(\varphi)$ is an action functional for the family of processes $(\tilde\vartheta_t^\gamma, t \le T)$ in $C[0,T;E^d]$, that is, for any $x \in E^d$, $s > 0$ the set $\Phi_x(s) = (\varphi\colon\ \varphi \in C[0,T;E^d], \varphi_0 = x, S_{0T}(\varphi) \le s)$ is a compact in $C[0,T;E^d]$, and for any r, s, $\delta > 0$ and $\vartheta_0 = \mathbf{x}$ the inequalities hold:

$$\inf_{x\in E^d,\,|y|\le r,\,\varphi\in\Phi_x(s)} \liminf_{\gamma\to 0} \gamma \ln \mathsf{P}(\rho_{0T}(\tilde\vartheta^\gamma,\varphi) < \delta) \ge -S_{0T}(\varphi), \tag{11.14}$$

$$\sup_{x\in E^d,\,|y|\le r} \limsup_{\gamma\to 0} \gamma \ln \mathsf{P}(\rho_{0T}(\tilde\vartheta^\gamma, \Phi_x(s)) > \delta) \le -s. \tag{11.15}$$

As in 11.1, the action functional is equal to zero on averaging paths $\bar\vartheta$. Indeed, by virtue of Jensen inequality

$$G(\vartheta,\alpha) = \lim_{n\to\infty} n^{-1}\ln \mathsf{E}\exp\left(\alpha\sum_{k=1}^{n} f(\vartheta, y_k^\vartheta)\right) \ge \lim_{n\to\infty} n^{-1}\mathsf{E}\sum_{k=1}^{n}\alpha f(\vartheta, y_k^\vartheta) = \alpha\bar f(\vartheta),$$

so

$$L(\varphi_t,\dot\varphi_t) = \sup_\alpha(\dot\varphi_t\alpha - G(\varphi_t,\alpha)) \le \sup_\alpha(\dot\varphi_t\alpha - \alpha\bar f(\varphi_t)) = 0$$

for $\dot\varphi_t = \bar f(\varphi_t)$, and, on the other hand, $L \ge 0$. Thus, $L(\varphi_t,\dot\varphi_t) = 0$ for $\dot\varphi_t = \bar f(\varphi_t)$.

From our assumptions and arguments from Ellis (1985) it follows that $S(\varphi) \ne 0$ for any other path. Indeed, Remark 8.1 implies that $L(\varphi_t,\cdot)$ is equal to zero in the unique point. Thus, $S(\varphi) = 0$ if and only if φ satisfies the equality $\dot\varphi_t = \bar f(\varphi_t)$.

Preliminary results

Let us denote $\tau_0 = 0$, $\tau = \inf(k \geq 0\colon |y_k^z| \leq R)$, $\gamma_n^0 = \tau_n + 1$, $\gamma_n^{i+1} = \vartheta_{\gamma_n^i}\tau + 1$, $\tau_{n+1} = \inf(\gamma_n^i\colon |y_{\gamma_n^i}^z| \leq R,\ i \geq 0)$, $n \geq 0$.

Lemma 11.1. Let conditions A, B and C be satisfied. Then for any c, $\nu > 0$ and $C > 1$ there exists such R_0 that for any $R \geq R_0$ and any x, z, y

$$\mathsf{E}_y \exp(c\tau_1) \leq C \exp(\nu(|y| - R)_+).$$

Proof is analogous to the proof of Lemma 8.1.

Let us denote $n(t) = \sup(n = 0, 1, 2, \ldots\colon \tau_n \leq t)$.

Lemma 11.2. Let conditions A, B and C be satisfied. Then for any $\delta > 0$, $C > 0$ for $R \geq R(C, \delta)$, $t \geq t(\delta, y)$ the following estimate holds:

$$\mathsf{P}(|t^{-1}n(t) - \nu_R| > \delta) \leq \exp(-Ct).$$

For any $\delta > 0$, $C > 0$, $\nu > 0$ for R large enough and any $t \geq t(\delta), y, x, z$

$$\mathsf{P}(|t^{-1}n(t) - \nu_R| > \delta) \leq \exp(-Ct)\exp(\nu(|y| - R + 1)_+).$$

Proof is analogous to the proof of Lemma 8.9.

It follows from Lemma 11.2 that

$$\lim_{R \to \infty} \nu_R = 1.$$

Now, let us denote $B_R = (y\colon |y| \leq R)$, $L_\infty(B_R)$ is the space of essentially bounded functions on B_R, $L_\infty^+(B_R)$ is the cone of non-negative functions from $L_\infty(B_R)$. For R fixed consider the semigroup of operators T_n^α, $n = 0, 1, 2, \ldots$ on $L_\infty(B_R)$ defined by the formula $T_n^\alpha g(y) = \mathsf{E}_y g(y_{\tau_n}^z) \exp(\sum_{k=0}^{\tau_n} \alpha f(x, y_k^z))$. Operators T_n^α are well-defined on $L_\infty(B_R)$ for any α and R large enough because of the existence of the transition density and by virtue of Lemma 11.1.

Lemma 11.3. Let conditions A, B and C be satisfied. Then operator T^α is 1-bounded, if $|\alpha| \leq a$, $R \geq R(a)$.

Proof is analogous to the proofs of Lemmas 8.3 and 8.4.

Lemma 11.4. Let conditions A, B and C be satisfied. Then the spectral radius $r(T^\alpha)$ is a simple eigen-value of T^α and its eigen-function e_R belongs to the cone $L_\infty^+(B_R)$. Moreover, function $r(T^\alpha)$ is convex in α and differentiable in α if $|\alpha| \leq a$, $R \geq R(a)$.

Proof is analogous to the proof of Lemma 8.6.

Lemma 11.5. Let $a > 0$, $\alpha \in E^l$, $|\alpha| \leq a$, and let conditions A, B and C be satisfied. Then there exists such $R_a > 0$ that for any $R \geq R_a$ there exists the limit

$$\mu_R(x, z, \alpha) = \lim n^{-1} \ln \mathsf{E}_y \exp\left(\sum_{k=0}^{\tau_n} \alpha f(x, y_k^z)\right).$$

Function $\mu_R(x, z, \alpha)$ is differentiable in α for $|\alpha| < a$, $R \geq R_a$, and convex in α. There exists such $C(R)$ that for any y and n

$$\left| n^{-1} \mathsf{E}_y \exp\left(\sum_{k=0}^{\tau_n} \alpha f(x, y_k^z)\right) - \mu_R(x, z, \alpha) \right| \leq C(R) n^{-1}.$$

Proof is analogous to the proof of Lemma 8.8.

Lemma 11.6. Let conditions A, B and C be satisfied. Then for any $a > 0$ for $R \geq R_a$ function μ_R is uniformly continuous in (x, z, α), $\alpha \leq a$.

Proof. Since function f is bounded then (11.2) implies the Lipschitz condition for μ_R in α with unique constant whenever function μ_R is well-defined. For any fixed a function μ_R is uniformly continuous in x by virtue of (11.2) and uniform continuity of f in x. It remains to check the uniform continuity of μ_R in z. However, it follows from continuity condition of a and b in the first variable by virtue of Lemma 11.1 and inequality of Lemma 11.5. Lemma 11.6 is proved.

Proof of Theorem 11.3. Let us prove the existence of the limit in (11.11). Fix $\alpha \in E^d$. For R large enough we have

$$n^{-1} \ln \mathsf{E} \exp\left(\alpha \sum_{k=1}^{n} f(\vartheta, Y_n^\vartheta)\right)$$

$$= \nu_R (\nu_R n)^{-1} \ln \mathsf{E}_y \exp\left(\alpha \left(\sum_{k=1}^{\tau_{[\nu_R n]}} + \sum_{k=\tau_{[\nu_R n]}+1}^{n}\right) f(\vartheta, Y_n^\vartheta)\right).$$

Let us estimate the expression under ln. Let $o < \delta < 1$. Then

$$\mathsf{E}_y \exp\left(\alpha \left(\sum_{k=1}^{\tau_{[\nu_R n]}} + \sum_{k=\tau_{[\nu_R n]}+1}^{n}\right) f(\vartheta, Y_n^\vartheta)\right) \leq \mathsf{E}_y \exp\left(\alpha \sum_{k=1}^{\tau_{[\nu_R n]}} f(\vartheta, Y_n^\vartheta)\right.$$

$$+ |\alpha| \|f\|_{\mathbf{B}} |n - \tau_{[\nu_R n]}| \Big) \mathbf{I}\left(|n - \tau_{[\nu_R n]}| \leq \delta n\right)$$

$$+ \exp\left(|\alpha| \|f\|_{\mathbf{B}} n\right) \mathsf{E}_y \mathbf{I}\left(|n - \tau_{[\nu_R n]}| > \delta n\right).$$

By virtue of Lemma 11.2 we have for R and n large enough

$$\mathsf{P}_y(|n - \tau_{[\nu_R n]}| > \delta n) \leq 2 \exp(-Cn),$$

where C may be chosen arbitrary large if R is large enough. Hence,

$$n^{-1} \ln \mathsf{E} \exp\left(\alpha \sum_{k=1}^{n} f(\vartheta, Y_n^\vartheta)\right) \leq 2 \exp\left(|\alpha| \|f\|_{\mathbf{B}} n - Cn\right)$$

$$+ n^{-1} \ln \mathsf{E} \exp\left(\alpha \sum_{k=1}^{[\nu_R n]} f(\vartheta, Y_n^\vartheta)\right.$$

$$\left. + |\alpha| \|f\|_{\mathbf{B}} \delta n\right).$$

So

$$n^{-1}\ln \mathsf{E}\exp\left(\alpha\sum_{k=1}^{n} f(\vartheta, Y_n^{\vartheta})\right)$$

$$\leq n^{-1}\ln\Bigg(\mathsf{E}_y \exp\left(\alpha\sum_{k=1}^{[\nu_R n]} f(\vartheta, Y_n^{\vartheta})\right)\exp\left(|\alpha|\|f\|_{\mathbf{B}}\delta n\right)$$

$$+2\exp\left(|\alpha|\|f\|_{\mathbf{B}} n - Cn\right)\Bigg).$$

By virtue of Lemma 11.5 for any $\rho > 0$ and for n large enough

$$\mu_R(x,z,\alpha) - \rho \leq [\nu_R n]^{-1}\ln \mathsf{E}_y \exp\left(\alpha\sum_{k=1}^{[\nu_R n]} f(x, y_n^z)\right) \leq \mu_R(x,z,\alpha) + \rho,$$

or, equivalently,

$$\begin{aligned}\exp\left((\mu_R(x,z,\alpha) - \rho)[\nu_R n]\right) &\leq [\nu_R n]^{-1}\ln \mathsf{E}_y \exp\left(\alpha\sum_{k=1}^{[\nu_R n]} f(x, y_n^z)\right)\\ &\leq \exp\left((\mu_R(x,z,\alpha) + \rho)[\nu_R n]\right).\end{aligned}$$

Since $\delta > 0$ and $C > 0$ may be chosen arbitrary then

$$\limsup_{n\to\infty} n^{-1}\ln T_n^{\alpha}1(y) \leq \liminf_{R\to\infty} \nu_R\mu_R(x,z,\alpha).$$

Similarly we obtain the lower bound

$$\liminf_{n\to\infty} n^{-1}\ln T_n^{\alpha}1(y) \geq \limsup_{R\to\infty} \nu_R\mu_R(x,z,\alpha).$$

This implies the existence of the limit

$$\lim_{n\to\infty} n^{-1}\ln \mathsf{E}\exp\left(\alpha\sum_{k=1}^{n} f(\vartheta, y_n^{\vartheta})\right) = \lim_{R\to\infty} \nu_R\mu_R(\vartheta,\vartheta,\alpha).$$

We have obtained (11.11) with $G(\alpha,\vartheta) = \lim_{R\to\infty}\nu_R\mu_R(\vartheta,\vartheta,\alpha)$. Inequalities (11.13) are deduced by the same considerations with the help of the second inequality of Lemma 11.2. Inequalities (11.13) and Lemma 11.6 imply the continuity of G in all variables. Theorem 11.3 is proved.

Proof of Theorem 11.4. (1) The compactness of the set $\Phi_x(s)$ follows from the properties of function L.

(2) Let us choose such small $\Delta > 0$ that $T/\Delta = m$ is the integer. Let ψ_t be a right-continuous step function with values in R^d, φ_t^{λ} is a smooth function with values in R^d such that $\rho_{0T}(\varphi,\varphi^{\lambda}) < \lambda$ and $\rho_{0T}(\varphi,\psi^{\lambda}) < \lambda$. Once more restriction for the choice of φ^{λ} will be done a bit later.

Since f and its derivatives are bounded then

$$(\omega\colon\ \rho_{0T}(\varphi,\vartheta^{\gamma}) < \delta) \supset (\omega\colon\ \rho_{0T}(\tilde{\vartheta}^{\gamma,\psi^{\lambda}},\varphi^{\lambda}) < \delta')$$

hence,

$$\mathsf{P}(\rho_{0T}(\varphi, \tilde{\vartheta}^{\gamma}) < \delta) \geq \mathsf{P}(\rho_{0T}(\tilde{\vartheta}^{\gamma,\psi^{\lambda}}, \varphi^{\lambda}) < \delta'),$$

if $\lambda = \lambda(\delta)$ and $\delta' = \delta'(\delta)$ are small enough; here

$$\tilde{\vartheta}^{\gamma,\psi^{\lambda}}_{(n+1)\gamma} = \tilde{\vartheta}^{\gamma,\psi^{\lambda}}_{n\gamma} + f(\psi^{\lambda}_{n\gamma}, Y_n), \qquad \tilde{\vartheta}^{\gamma,\psi^{\lambda}}_0 = \vartheta_0.$$

Let us denote

$$\tilde{S}_{0T}(\varphi) = \tilde{S}^{\psi^{\lambda}}_{0T}(\varphi) = \int_0^T L(\psi^{\lambda}_{n\gamma}, \dot{\varphi}^{\lambda}_s)\,\mathrm{d}s.$$

(3) Functional $\tilde{S}$ is lower semicontinuous in (φ, ψ) in uniform topology. It is proved as in Freidlin and Wentzell (1984).

(4) Let us prove that for $\nu > 0$ small enough

$$\mathsf{P}(\rho_{0T}(\tilde{\vartheta}^{\gamma,\psi^{\lambda}}, \varphi^{\lambda}) < \delta') \geq \exp(-\gamma^{-1}(\tilde{S}^{\psi^{\lambda}}_{0T}(\varphi) + \nu)).$$

Denote

$$\varphi^{\lambda,\Delta} = (\varphi^{\lambda}_{\Delta}, \ldots, \varphi^{\lambda}_{m\Delta}), \qquad \Delta^1_R = \gamma\tau^1_{[\Delta/\gamma]+1},$$

where $\tau^1_0 = 0$, $\tau^1 = \inf(k \geq 0\colon |y^{\vartheta_0}_k| \leq R)$, $\gamma^{1,n}_0 = \tau^1_n + 1$, $\gamma^{1,n}_{i+1} = \vartheta_{\gamma^{1,n}_i} + 1$ (ϑ is a shift operator), $\tau^1_{n+1} = \inf(\gamma^{1,n}_i\colon |y^{\vartheta_0}_{\gamma^{1,n}_i}| \leq R,\ i = 0, 1, \ldots)$, $n \geq 0$. Denote also

$$\Delta^2_R = \gamma\tau^2_{[\Delta/\gamma]+1},$$

where $\tau^2_0 = \Delta^1_R$, $\tau^2 = \inf(k \geq 0\colon |y^{\vartheta^1}_k| \leq R)$, $\vartheta^i = \tilde{\vartheta}_{\Delta^i_R}$, $\hat{\vartheta}^{\psi^{\lambda}}_{\Delta^i_R} = \vartheta^i$, $\hat{\vartheta}^{\psi^{\lambda}}_{(n+1)\gamma} = \hat{\vartheta}^{\psi^{\lambda}}_{n\gamma} + \gamma f(\psi^{\lambda}_{n\gamma}, y^{\vartheta^i}_n)$, $\Delta^i_R \leq n\gamma < \Delta^{i+1}_R$, $\gamma^{2,n}_0 = \tau^2_n + 1$, $\gamma^{2,n}_{i+1} = \vartheta_{\gamma^{2,n}_i} + 1$, $\tau^{2,n}_{i+1} = \vartheta_{\gamma^{2,n}_i}\tau^2 + 1$, $\tau^2_{n+1} = \inf(\gamma^{2,n}_i\colon |y^{\vartheta_1}_{\gamma^{2,n}_i}| \leq R,\ i = 0, 1, \ldots)$, $n \geq 0$, etc. Note that always $\Delta^i_R \geq T_i = iT/m$.

(5) We have for δ'', δ''' small enough

$$\begin{aligned}\mathsf{P}(\rho_{0T}(\tilde{\vartheta}^{\gamma,\psi^{\lambda}}, \varphi^{\lambda}) < \delta') \geq\ & \mathsf{E} \prod_{1 \leq k \leq m} \mathrm{I}(|\tilde{\vartheta}^{\gamma,\psi^{\lambda}}_{\Delta^k_R} - \varphi^{\lambda}_{k\Delta}| < km^{-1}\delta'') \\ & \times \prod_{1 \leq k \leq m} \mathrm{I}(|\Delta^k_R - k\Delta| \leq \delta'''\Delta) \\ & - \mathsf{P}\left(\bigcup_{1 \leq k \leq m} |\Delta^k_R - k\Delta| > \delta'''\Delta\right).\end{aligned}$$

By virtue of Lemma 11.2 the probability $\mathsf{P}(\bigcup_k |\Delta^k_R - k\Delta| > \delta'''\Delta)$ may be done less than $\exp(-C\gamma^{-1})$ with any $C > 0$ if R is large enough. Let us estimate the probability $\mathsf{P}(|\tilde{\vartheta}^{\gamma,\psi^{\lambda}}_{\Delta^m_R} - \varphi^{\lambda}_T| < \delta'' | \tilde{\vartheta}^{\gamma,\psi^{\lambda}}_{\Delta^{m-1}_R} = \vartheta^{m-1})$ if $|\vartheta^{m-1} - \varphi^{\lambda}_{T_{m-1}}| < \delta'_{m-1}$.

(6) Let $\alpha \in R^d$, $|\alpha| < a$ (number $a > 0$ will be chosen later). Let us denote for such α

$$\tilde{h}_{\gamma,R}(\alpha) = \gamma \ln \mathsf{E}(\exp(\gamma^{-1}\alpha(\tilde{\vartheta}^{\gamma,\psi^{\lambda}}_{\Delta^m_R} - \tilde{\vartheta}^{\gamma,\psi^{\lambda}}_{\Delta^{m-1}_R}))|F_{\Delta^{m-1}_R})$$

$$= \gamma \ln \mathsf{E}\left(\exp\left(\sum_{\Delta_R^{m-1}\le k\gamma<\Delta_R^m} \alpha f(\psi_{k\gamma}^\lambda, Y_k)\right) |F_{\Delta_R^{m-1}}\right),$$

$$h_{\gamma,R}(\alpha) = \gamma \ln \mathsf{E}(\exp(\gamma^{-1}\alpha(\hat{\vartheta}_{\Delta_R^m}^{\gamma,\psi^\lambda} - \hat{\vartheta}_{\Delta_R^{m-1}}^{\gamma,\psi^\lambda}))|F_{\Delta_R^{m-1}})$$

$$= \gamma \ln \mathsf{E}\left(\exp\left(\sum_{\Delta_R^{m-1}\le k\gamma<\Delta_R^m} \alpha f(\psi_{k\gamma}^\lambda, y_{k\gamma}^{\psi^\lambda})\right) |F_{\Delta_R^{m-1}}\right),$$

and let $\tilde{h}_{\gamma,R}(\alpha) = +\infty$, $h_{\gamma,R}(\alpha) = +\infty$ for all other α.

For $|\alpha| < a$ and R large enough there exists the limit

$$h_R(\alpha) = \lim_{\gamma\to 0} h_{\gamma,R}(\alpha) = \int_{\Delta_R^{m-1}}^{\Delta_R^m} \Delta G_R(\vartheta^{m-1}, \alpha)\,\mathrm{d}s,$$

where we define on the set $(\psi_{\Delta_R^{m-1}}^\lambda = \vartheta^{m-1})$

$$G_R(\vartheta^{m-1}, \alpha) = \lim_{\gamma\to 0} \gamma\Delta^{-1} \ln \mathsf{E}\left(\exp\left(\alpha \sum_{\Delta_R^{m-1}\le k\gamma<\Delta_R^m} f(\vartheta^{m-1}, y_{k\gamma}^{\psi^\lambda})\right) |F_{\Delta_R^{m-1}}\right)$$

(recall that $|\alpha| < a$). The existence of both of these limits follows from Lemma 11.5. Moreover, for any Δ small enough we have on the set $(\tilde{\vartheta}_{\Delta_R^{m-1}}^\gamma = \vartheta^{m-1})$

$$\begin{aligned} G_R(\vartheta^{m-1}, \alpha) - \delta\Delta &\le \liminf_{\gamma\to 0} \gamma\Delta^{-1} \ln \mathsf{E}\left(\exp\left(\alpha \sum_{\Delta_R^{m-1}\le k\gamma<\Delta_R^m} f(\vartheta^{m-1}, Y_k)\right) |F_{\Delta_R^{m-1}}\right) \\ &\le \limsup_{\gamma\to 0} \gamma\Delta^{-1} \ln \mathsf{E}\left(\exp\left(\alpha \sum_{\Delta_R^{m-1}\le k\gamma<\Delta_R^m} f(\vartheta^{m-1}, Y_k)\right) |F_{\Delta_R^{m-1}}\right) \\ &\le G_R(\vartheta^{m-1}, \alpha) + \delta\Delta \end{aligned}$$

uniformly w.r.t. $|y| \le R$.

(7) Consider the following measures on $F_{\Delta_R^{m-1}}$: $\mathrm{d}\mathsf{P}_R^{\gamma,\alpha} = \exp(\gamma^{-1}(\alpha(\hat{\vartheta}_{\Delta_R^m}^{\gamma,\psi^\lambda} - \hat{\vartheta}_{\Delta_R^{m-1}}^{\gamma,\psi^\lambda}) - h_{\gamma,R}(\alpha))\,\mathrm{d}\mathsf{P}$ and $\mathrm{d}\tilde{\mathsf{P}}_R^{\gamma,\alpha} = \exp(\gamma^{-1}(\alpha(\tilde{\vartheta}_{\Delta_R^m}^{\gamma,\psi^\lambda} - \tilde{\vartheta}_{\Delta_R^{m-1}}^{\gamma,\psi^\lambda}) - h_{\gamma,R}(\alpha))\,\mathrm{d}\mathsf{P}$. Both of them are probabilistic ones (here $|\alpha| < a$). We have on the set $(\tilde{\vartheta}_{\Delta_R^{m-1}}^\gamma = \vartheta^{m-1})$

$$\begin{aligned} \mathsf{P}(|\tilde{\vartheta}_{\Delta_R^m}^{\gamma,\psi^\lambda} - \varphi_T^\lambda| < \delta'' |F_{\Delta_R^{m-1}}) &= \tilde{\mathsf{E}}_R^{\gamma,\alpha}(\mathrm{I}(|\tilde{\vartheta}_{\Delta_R^m}^{\gamma,\psi^\lambda} - \varphi_T^\lambda| < \delta'') \\ &\quad \times \exp(-\gamma^{-1}(\alpha(\tilde{\vartheta}_{\Delta_R^m}^{\gamma,\psi^\lambda} - \tilde{\vartheta}_{\Delta_R^{m-1}}^{\gamma,\psi^\lambda}) + \tilde{h}_{\gamma,R}(\alpha)))|F_{\Delta_R^{m-1}}). \end{aligned}$$

Let us explain how to choose α here. Denote by $L_R^\gamma(\vartheta^{m-1}, \beta)$ the Fenchel–Legendre transformation of $\Delta^{-1}h_{\gamma,R}(\alpha)$. Function $h_{\gamma,R}$ is differentiable in α for any $|\alpha| < a$ for $R \ge R_a$. Hence (see Chapter 8, proof of Theorem 8.9) there exists such $\alpha_0 = \alpha_0(R)$ that $|\alpha_0| < a$ and

$$L_R^\gamma(\vartheta^{m-1}, \beta) > L_R^\gamma(\vartheta^{m-1}, (\varphi_T^\lambda - \varphi_{T_{m-1}}^\lambda)/\Delta) + \alpha_0(\varphi_T^\lambda - \varphi_{T_{m-1}}^\lambda)/\Delta,$$

$$\beta \neq (\varphi_T^\lambda - \varphi_{T_{m-1}}^\lambda)/\Delta.$$

Then

$$\begin{aligned}\Delta^{-1}h_{\gamma,R}(\alpha_0) &= \sup(\alpha_0\beta - L_R^\gamma(\vartheta^{m-1},\beta)) \\ &= \alpha_0(\varphi_T^\lambda - \varphi_{T_{m-1}}^\lambda)/\Delta - L_R^\gamma(\vartheta^{m-1},(\varphi_T^\lambda - \varphi_{T_{m-1}}^\lambda)/\Delta).\end{aligned}$$

Denote $\hat{F}_{m-1} = F_{\Delta_R^{m-1}}$. Then we have for $\alpha = \alpha_0$ on $(\rho_{0,\Delta_R^{m-1}}(\tilde{\vartheta}^{\gamma,\psi^\lambda},\varphi^\lambda) < \delta')$)

$$\begin{aligned}&\tilde{\mathsf{E}}_R^{\gamma,\alpha}[\mathrm{I}(|\tilde{\vartheta}^{\gamma,\psi^\lambda}_{\Delta_R^m} - \varphi_T^\lambda| < \delta'')\exp(-\gamma^{-1}(\alpha(\tilde{\vartheta}^{\gamma,\psi^\lambda}_{\Delta_R^m} - \tilde{\vartheta}^{\gamma,\psi^\lambda}_{\Delta_R^{m-1}}) + \tilde{h}_{\gamma,R}(\alpha)))|\hat{F}_{m-1}] \\ &\quad\geq \exp(-\gamma^{-1}(\alpha_0(\tilde{\vartheta}^{\gamma,\psi^\lambda}_{\Delta_R^m} - \vartheta^{m-1}) - \Delta G_R(\vartheta^{m-1},\alpha_0) \\ &\quad\quad + \Delta G_R(\vartheta^{m-1},\alpha_0) + h_{\gamma,R}(\alpha_0) + \nu/2)) \\ &\quad\quad \times \tilde{\mathsf{E}}_R^{\gamma,\alpha}[\mathrm{I}(|\tilde{\vartheta}^{\gamma,\psi^\lambda}_{\Delta_R^m} - \varphi_T^\lambda| < \delta'')|\hat{F}_{m-1}] \\ &\quad\geq \exp(-\gamma^{-1}\Delta(L_R(\vartheta^{m-1},(\varphi_T^\lambda - \varphi_{T_{m-1}}^\lambda)/\Delta) + \nu)) \\ &\quad\quad \times \mathsf{E}_R^{\gamma,\alpha}[\mathrm{I}(|\hat{\vartheta}^{\gamma,\psi^\lambda}_{\Delta_R^m} - \varphi_T^\lambda| < \delta'')|\hat{F}_{m-1}] - \exp(-C\gamma^{-1}).\end{aligned}$$

(8) We have on the set $(|\Delta_R^{m-1} - T_{m-1}| \leq \delta''\Delta)$

$$\mathsf{E}_R^{\gamma,\alpha}[\mathrm{I}(|\hat{\vartheta}^{\gamma,\psi^\lambda}_{\Delta_R^m} - \varphi_T^\lambda| \geq \delta'')|\hat{F}_{m-1}] \leq \exp(-\gamma^{-1}\Delta s_m)$$

for some $s_m > 0$. This estimate follows from the upper bound for the family of measures $\mathsf{P}_R^{\gamma,\alpha}$ in the same way as in Gärtner's theorems (see Chapter 8).

(9) The latter inequality implies

$$\begin{aligned}&\mathsf{P}[|\tilde{\vartheta}^{\gamma,\psi^\lambda}_{\Delta_R^m} - \varphi_T^\lambda| < \delta''|\hat{F}_{m-1}] \\ &\quad\geq \exp(-\gamma^{-1}\Delta(L_R(\psi_{T_{m-1}}^\lambda,(\varphi_T^\lambda - \varphi_{T_{m-1}}^\lambda)/\Delta) + \nu) \\ &\quad\quad \times[1 - \exp(-\gamma^{-1}s_m\Delta)] - \exp(-C\gamma^{-1}).\end{aligned}$$

(10) Similarly we obtain on the set $(|\Delta_R^{m-2} - T_{m-2}| \leq \delta'''\Delta)$

$$\begin{aligned}&\mathsf{E}[\mathrm{I}(|\tilde{\vartheta}^{\gamma,\psi^\lambda}_{\Delta_R^{m-1}} - \varphi_{T_{m-1}}^\lambda| < \delta'')|\hat{F}_{m-2}] \\ &\quad\geq \exp(-\gamma^{-1}\Delta(L_R(\psi_{T_{m-2}}^\lambda,(\varphi_{T_{m-1}}^\lambda - \varphi_{T_{m-2}}^\lambda)/\Delta) + \nu) \\ &\quad\quad - \exp(-C\gamma^{-1}).\end{aligned}$$

By induction we obtain on $(|\Delta^{m-k} - T_{m-k}| \leq \delta'''\Delta)$

$$\begin{aligned}&\mathsf{E}[\mathrm{I}(|\tilde{\vartheta}^{\gamma,\psi^\lambda}_{\Delta_R^{m-1}} - \varphi_T^\lambda| < \delta'')\ldots\mathrm{I}(|\tilde{\vartheta}^{\gamma,\psi^\lambda}_{\Delta_R^{m-k+1}} - \varphi_{T_{m-k+1}}^\lambda| < \delta'')|\hat{F}_{m-k}] \\ &\quad\geq \exp\left(-\gamma^{-1}\Delta\sum_{i=1}^{k}(L_R(\psi_{T_{m-i}}^\lambda,(\varphi_{T_{m-i+1}}^\lambda - \varphi_{T_{m-i}}^\lambda)/\Delta) + \nu)\right) \\ &\quad\quad - \exp(-C\gamma^{-1}), \qquad k = 1,\ldots,m-1.\end{aligned}$$

If we take $R > |y|$ then this estimate holds for $k = m$ also. Then we obtain

$$\mathsf{P}(|\tilde{\vartheta}^{\gamma,\psi^\lambda}_{\Delta_R^{m-1}} - \varphi^\lambda_T| < \delta'', \dots, |\tilde{\vartheta}^{\gamma,\psi^\lambda}_{\Delta_R^1} - \varphi^\lambda_{T_1}| < \delta'')$$
$$\geq \exp\left(-\gamma^{-1}\Delta\sum_{i=1}^{m}(L_R(\psi^\lambda_{T_{m-i}}, (\varphi^\lambda_{T_{m-i+1}} - \varphi^\lambda_{T_{m-i}})/\Delta) + \nu)\right) - \exp(-C\gamma^{-1})$$
$$= \exp\left(-\gamma^{-1}\Delta\sum_{i=0}^{m-1}(L_R(\psi^\lambda_{T_i}, (\varphi^\lambda_{T_{i+1}} - \varphi^\lambda_{T_i})/\Delta) + \nu)\right) - \exp(-C\gamma^{-1}).$$

(11) Let us derive now the estimate

$$\mathsf{P}(\rho(\tilde{\vartheta}^{\gamma,\psi^\lambda}, \varphi^\lambda) < \delta'') \geq \exp(-\gamma^{-1}(\tilde{S}^{\psi^\lambda,R}_{0T}(\varphi^\lambda) + \nu)).$$

It is proved similarly to the proof of Lemma 7.5.1 from Freidlin and Wentzell (1984). Indeed, since L_R is convex then we have

$$\Delta\sum L_R\left(\psi_{k\Delta}, \Delta^{-1}\int_{k\Delta}^{(k+1)\Delta}\dot{\varphi}_s\,ds\right) \leq \sum\int_{k\Delta}^{(k+1)\Delta} L_R(\psi_{k\Delta}, \dot{\varphi}_s)\,ds = \int_0^T L_R(\psi_s, \dot{\varphi}_s)\,ds.$$

(12) By virtue of Lemma 7.5.2 from Freidlin and Wentzell (1984) and differentiability of G_R in α for $|\alpha| < a$ we deduce that for any sequence of step functions ψ^n which converges to φ uniformly there exists such sequence $\varphi^n \in C[0,T;E^l]$ which also converges uniformly to φ that

$$\limsup_n \int_0^T L_R(\psi^n_s, \dot{\varphi}^n_s)\,ds \leq S^R_{0T}(\varphi),$$

if $|\dot{\varphi}_s| \leq a$ for any s. If this inequality fails for some s but $S^R_{0T}(\varphi) < \infty$ then we may simply take $\varphi^a_t = \int_0^t \dot{\varphi}_s \mathrm{I}(|\dot{\varphi}_s| \leq a)\,ds$. For a large enough we obtain $\rho_{0T}(\varphi, \varphi^a) < \delta/2$ and $S^R_{0T}(\varphi^a) \leq S^R_{0T}(\varphi) + \delta/2$ since the integral $\int_0^T L_R(\varphi_s, \dot{\varphi}_s)\,ds$ converges.

(13) Finally, from (12) and (2) we obtain

$$\mathsf{P}(\rho_{0T}(\tilde{\vartheta}^\gamma, \varphi) < \delta) \geq \exp(-\gamma^{-1}(S^R_{0T}(\varphi) + \nu)) - \exp(-C\gamma^{-1}), \qquad \gamma \to 0.$$

(14) Similarly to Chapter 8 (see the proof of Theorem 8.9) we obtain

$$S^R_{0T}(\varphi) \leq S_{0T}(\varphi) + \nu, \qquad \nu > 0$$

as $R \to \infty$. Thus, with $C > S_{0T}(\varphi)$ we get

$$\mathsf{P}(\rho_{0T}(\tilde{\vartheta}^\gamma, \varphi) < \delta) \geq \exp(-\gamma^{-1}(S_{0T}(\varphi) + \nu)), \qquad \gamma \to 0.$$

(15) The second inequality is proved as in Theorem 11.2 absolutely similarly to the correspondent part of Theorem 7.4.1 from Freidlin and Wentzell (1984). Theorem 11.4 is proved.

Comments

Results of this chapter are essentially extended variants of theorems from Freidlin (1978), see also Freidlin and Wentzell (1984). The main extension is noncompactness of the state space of the fast process. Compactness of this state space was important in Freidlin (1978) and Freidlin and Wentzell (1984).

Another difference is of course discrete time; however, more important is the dependence of the diffusion coefficient b on slow process (in Freidlin's theorems such dependence was not allowed). In Freidlin's case it is possible to use Girsanov transformation of measure which reduces the problem to independent fast process. In our case it is impossible. Nevertheless we may prove the theorem after changing the sequence of steps in the original Freidlin's proof.

See Dupuis (1988) and Dupuis and Kushner (1987, 1989) once more about the importance of differentiability of the limit considered for stochastic approximation theory.

The exposition is based on the papers of Veretennikov (1990, 1991b) for continuous time; similar results were announced in Veretennikov (1989b) but proofs are simplified essentially.

The 'next chapters' must be devoted to asymptotics of hitting times, quasipotentials and so on, similar to Freidlin (1978) and Freidlin and Wentzell (1984), but it is the material for future papers.

Bibliography

De Acosta, A., Ney, P. and Nummelin, E. Large deviation lower bounds for general sequences of random variables. In: *Random Walks, Brownian Motion and Interacting Particle Systems* (R. Durret and H. Kesten, eds). Birkhäuser, Boston, 1991, pp. 215–221.

Azencott, R. Grandes deviations et applications. *Lecture Notes Math.* (1980) **774**, 1–176.

Bahadur, R. R. and Zabell, S. L. Large deviations of the sample mean in general vector spaces. *Ann. Probab.* (1979) **7**, 587–621.

Baldi, P. and Chaleyat-Maurel, M. An extansion of Ventsel–Freidlin estimates. *Lecture Notes Math.* (1988) **1316**, 305–327.

Benveniste, A., Metivier, M. and Priouret, P. *Algorithmes Adaptifs et Approximations Stochastiques. Théorie et Applications a l'Identification, au Traitment du Signal et a Reconnaissance des Formes*. Masson, Paris, 1987.

Benveniste, A., Metivier, M. and Priouret, P. *Adaptive Algorithms and Stochastic Approximation.* Springer, Berlin, 1990. Applications of Math. (Transl. from French).

Berbee, H. and Bradley, R. C. A limitation of Markov representation for stationary processes. *Stoch. Proc. Appl.* (1984) **18**, 33–45.

Billingsley, P. *Convergence of Probability Measures*. Wiley, 1969.

Bloom, J. R., Hanson, D. L. and Koopman, L. H. On the strong law of large numbers for a class of stochastic processes. *Z. Wahrsch. Verv. Gebiete* (1963) **2**, 1–11.

Borovkov, A. A. Boundary-value problems for random walks and large deviations in function spaces. *Theory Probab. Appl.* (1967) **12**, 575–595.

Bradley, R. C. On the φ-mixing condition for stationary random sequences. *Duke Math. J.* (1980) **47**, 421–433.

Bradley, R. C. On the ψ-mixing condition for stationary random sequences. *Trans. Amer. Math. Soc.* (1983) **276**, 55–66.

Chernoff, H. A measure of asymptotic efficiency for tests of a hypothesis based on the sum of observations. *Ann. Math. Statist.* (1952) **23**, 493–507.

Chiyonobu, T. and Kusuoka, S. The large deviations principle for hypermixing processes. *Theory Probab. Rel. Fields* (1988) **78**, 627–649.

Cramér, H. Sur un nouveau théorème-limite de la thèorie des probabilités. *Actualités Scientifiques et Industrielles* (1938) **736**, 5–23.

Davydov, Yu. A. Mixing conditions for Markov chains. *Teoria Veroyatnostey i Primen* (1973) **18**, 321–338 (in Russian).

Deuschel, J.-D. and Stroock, D.W. *Large Deviations.* Academic Press, New York, 1989.

Dobrushin, R. L. Markov processes with a large number of locally interacting components: existence of a limit process and its ergodicity. *Problems of Information Transmission* (1971) **7**, No. 2, 149–164.

Donsker, M. D. and Varadhan, S. R. S. Asymptotic evaluation of certain Markov process expectations for large time, I. *Comm. Pure Appl. Math.* (1975a) **28**, 1–49.

Donsker, M. D. and Varadhan, S. R. S. Asymptotic evaluation of certain Markov process expectations for large time, II. *Comm. Pure Appl. Math.* (1975b) **28**, 279–301.

Donsker, M. D. and Varadhan, S. R. S. Asymptotic evaluation of certain Markov process expectations for large time, III. *Comm. Pure Appl. Math.* (1977) **29**, 389–461.

Donsker, M. D. and Varadhan, S. R. S. Asymptotic evaluation of certain Markov process expectations for large time, IV. *Comm. Pure Appl. Math.* (1983) **36**, 183–212.

Doob, J. L. *Stochastic Processes.* Wiley, New York; Chapman and Hall, London, 1953.

Dupuis, P. Large deviations analysis of some recursive algorithms with state dependent noise. *Ann. Probab.* (1988) **16**, 1509–1536.

Dupuis, P. and Kushner, H. J. Asymptotic behaviour of constrained stochastic approximation via the theory of large deviations. *Theory Probab. Rel. Fields* (1987) **75**, 223–244.

Dupuis, P. and Kushner, H. J. Stochastic approximation and large deviations: upper bounds and w.p.1 convergence. *SIAM J. Control and Optimization* (1989) **27**, 1108–1135.

Ellis, R. S. *Entropy, Large Deviations and Statistical Mechanics. Grundlehren der Mathematischen Wissenschaften*, vol. 271. Springer, New York, 1985.

Freidlin, M. I. Averaging principle and large deviations. *Uspekhi Mat. Nauk* (1978) **33**, No. 5, 107–160 (in Russian).

Freidlin, M. I. and Wentzell, A. D. *Random Perturbations of Dynamical Systems.* Springer-Verlag, New York, 1984.

Gartnër, J. On large deviations from the unvariant measure. *Theory Probab. Rel. Fields* (1977) **22**, 24–39.

Gutland, N. J. An extansion of the Ventcel–Freidlin large deviation principle. *Stochastics* (1988) **24**, 121–149.

Ibragimov, I. A. Spectral functions of certain classes of Gaussian stationary processes. *Dokl. Acad. Sci. USSR* (1961) **137** 1046–1048 (in Russian).

Ibragimov, I. A. Stationary Gaussian sequences that satisfy strong mixing condition. *Dokl. Acad. Sci. USSR* (1962) **147**, 1282–1284 (in Russian).

Ibragimov, I. A. and Linnik, Yu. V. *Independent and Stationary Sequences of Random Values.* Groningen, Noordhoff, 1971 (1965, in Russian).

Iscoe, I., Ney, P. and Nummelin, E. Large deviations of uniformly recurrent Markov additive processes. *Adv. Appl. Math.* (1985) **6**, 373–412.

Kato, T. *Perturbation Theory for Linear Operators.* Springer, Berlin, 1976. 2nd ed.

Khasminskii, R. Z. On stochastic processes defined by differential equations with a small parameter. *Teoria Veroyatnostey i Primen.* (1966) **11**, 211.

Kifer, Yu. A discrete-time version of the Wentzell–Freidlin theory. *Ann. Probab.* (1990) **6**, 68–78.

Kolmogorov, A. N. Local limit theorem for classical Markov chains. *Izvestia AN SSSR, Ser. Mat.* (1949) **13**, 281–300 (in Russian).

Kolmogorov, A. N. and Rozanov, Yu. A. On strong mixing conditions for Gaussian stationary process. *Teoria Veroyatnostey i Primen.* (1960) **5**, 222–227 (in Russian).

Krasnoselskii, M. A., Lifshitz, E. A. and Sobolev, A. V. *Positive Linear Systems.* Helderman, Berlin, 1989.

Krylov, N. V. and Safonov, M. V. A certain property of solutions of parabolic equations with measurable coefficients. *Math. USSR Izvestiya* (1981) **16**, 151–164.

Kullback, S. and Leibler, R. A. On information and sufficiency. *Ann. Math. Statist.* (1951) **22**, 79–86.

Kushner, H. J. and Clark, D. S. *Stochastic Approximation Methods for Constrained and Unconstrained Systems.* Springer, Berlin, 1978.

Lanford, O. E. Entropy and equilibrium states in classical statistical mechanics. In: *Statistical Mechanics and Mathematical Problems* (A. Lenard, ed.). Lectures Notes in Physics, **20**. Springer, Berlin, 1973, pp. 1–113.

Ljung, L. *IEEE Transactions on Automatic Control* (1977) **AC-22**, No. 4, 551–575.

Liptser, R. Sh. and Shiryaev, A. N. *Statistics of Random Processes. I. General Theory*. Springer, New York, 1977.

Liptser, R. Sh. and Shiryaev, A. N. *Theory of Martingales*. Kluwer Academic Publishers, Dordrecht, 1989.

Liptser, R. Sh., Pukel'sheim, F. and Shiryaev, A. N. Necessary and sufficient conditions for contiguity and entire asymptotic separation of probability measures. *Russian Math. Survays* (1982) **32**, 107–136.

Liptser, R. Sh. and Pukhalskii, A. A. Limits theorems on large deviations for semimartingales. To be published in: *Stochastic and Stochastic Reports*.

Mogulskii, A. A. Large deviations for trajectories of multidimensional random walks. *Theory Probab. Appl.* (1976) **21**, 309–323.

Narita, K. Large deviation principle for diffusion processes. *Tsukuba J. Math.* (1988) **12**, 211–229.

Ney, P. and Nummelin, E. Markov additive processes. I. Eigenvalue properties and limit theorems. II. Large deviations. *Ann. Probab.* (1987) **15**, 561–592; 593–609.

Nummelin, E. *General Irreducible Markov Chains and Non-negative Operators*. Cambridge University Press, Cambridge, 1984.

Olla, S. Large deviations for almost Markovian processes. *Theory Probab. Rel. Fields* (1987) **76**, 395–409.

Orey, S. and Pelikan, S. Large deviation principles for stationary processes. *Ann. Probab.* (1988) **16**, 1481–1495.

Petrov, V. V. *Sums of Independent Random Variables*. Springer, Berlin, 1975.

Pukhalskii, A. A. On functional principle of large deviations. In: *New Trends in Probability and Statistics* (V. Sazonov and T. Shervashidze, eds). VSP/Mokslas, 1991, pp. 198–219.

Rockafellar, R. T. *Convex Analysis*. Princeton University Press, Princeton, 1970.

Rozanov, Yu. A. *Stationary Random Processes*. Holden-Day, San Francisco, 1967.

Ruelle, D. Correlation functionals. *J. Math. Phys.* (1965) **6**, 201–220.

Ruelle, D. A variational formulation of equilibrium statistical mechanics and the Gibbs phase rule. *Comm. Math. Phys.* (1967) **5**, 324–329.

Saulis, L. and Statulevicius, L. *Limit Theorems for Large Deviations*. Mokslas, Vilnius, 1989.

Sanov, I. N. On the probability of large deviations of random variables, *Math. Sb.* (1957) **42**, 11–44. (English translation in Selected Translations in Mathematical Statistics and Probability I (1961), pp. 213–244)

Schilder, M. Some asymptotics formula for Wiener integrals, *Trans. Amer. Math. Soc.* (1966) **125**, 63–85.

Stroock, D. W. *An Introduction to the Theory of Large Deviations*. Springer, New York, 1966.

Takahashi, Y. Entropy function (free energy) for dynamical systems and their random perturbations. In: *Proceedings Taniguchi Symposium on Stochastic Analysis at Katata and Kyoto* (K. Itô, ed.). Kinokuniya and North Holland, Tokyo, 1982.

Varadhan, S. R. S. Asymptotic probabilities and differential equations. *Comm. Pure Appl. Math.* (1966) **19**, 261–286.

Varadhan, S. R. S. *Large Deviations and Applications*. SIAM, Philadelphia, Pennsylvania, 1984.

Vaserstein, L. N. Markov processes over denumerable products of spaces, describing large systems of automata. *Problems of Information Transmission* (1969) **5**, 47–52.

Veretennikov, A. Yu. Bounds for the mixing rate in the theory of stochastic-equations. *Theory Probab. Appl.* (1987) **32**, 273–281.

Veretennikov, A. Yu. *On the Theory of Large Deviations and on the Spectral Radius of Integral Positive Operators. Statistics and Control of Random Processes.* Nauka, Moscow, 1989a.

Veretennikov, A. Yu. Optimal stabilization of controllable random processes with fast oscillating noise. *Aut. Remote Control* (1989b) **50**, 1061–1065.

Veretennikov, A. Yu. On large deviations in averaging principle for stochastic differential equations with periodic coefficients. 1. In: *Probability Theory and Mathematical Statistics.* Proc. Fifth Vilnius Conf. (1989). (B. Grigelionis *et al*, eds). Mokslas and Utrecht, Vilnius, Lithuania; VSP, The Netherlands VSP, (1990) **2**, 542–551.

Veretennikov, A. Yu. On the estimates of the mixing rate for Markov processes. *Lietuvos Matematikos Rinkinis* (1991a) **31**, No. 1, 40–49 (in Russian).

Veretennikov, A. Yu. On large deviations in averaging principle for stochastic differential equations with periodic coefficients. 2. *Math. USSR Izveztia* (1991b) (to appear) [in Russian: *Izvestia AN SSSR, Ser Mat.* (1991), **55**, 691–715].

Veretennikov, A. Yu. and Gulinsky, O. V. Rate of mixing and the averaging principle for stochastic recursive procedures. *Aut. Remote Control* (1990) **51**, 779–788.

Wentzell, A. D. and Freidlin M. I. On small pertubations of dynamical systems. *Russian Math. Surveys* (1970) **25**, 1–55.

Wentzell, A. D. and Freidlin M. I. Some problems concerning stability under small random perturbations. *Theory Prob. Appl.* (1972) **17**, 269–283.

Bei Fragen zur Produktsicherheit wenden Sie sich bitte an:
If you have any questions regarding product safety,
please contact:

Walter de Gruyter GmbH
Genthiner Straße 13
10785 Berlin
productsafety@degruyterbrill.com